# SpringerBriefs in Biology

More information about this series at http://www.springer.com/series/10121

Andrés Moya

# The Calculus of Life

## Towards a Theory of Life

 Springer

Andrés Moya
University of Valencia
Valencia
Spain

and

Valencian Region Foundation for the
    Promotion of Health and Biomedical
    Research
Valencia
Spain

ISSN 2192-2179                    ISSN 2192-2187   (electronic)
SpringerBriefs in Biology
ISBN 978-3-319-16969-9        ISBN 978-3-319-16970-5   (eBook)
DOI 10.1007/978-3-319-16970-5

Library of Congress Control Number: 2015936291

Springer Cham Heidelberg New York Dordrecht London

Printed on acid-free paper

Springer International Publishing AG Switzerland is part of Springer Science+Business Media
(www.springer.com)

# Contents

# Introduction

This work hinges on a supposed contradiction between the title and contents, and this requires an immediate explication. Ever since I thought up the project, some time ago now, I have believed that the work should be called *The Calculus of Life*. This book deals with Biology and theoretical approaches to Biology. However, these approaches are not quantitative, and they do not set out to create models for biological phenomena applying the Mathematics used to develop models in other sciences. This is especially the case with Calculus, which was to emerge with Newton and Leibniz, and has gone hand in hand with the development of physical sciences. In order to meet the proposed objectives, it would probably have seemed more natural to name the book *The Logic of Life*, because the theoretical developments of this work are more qualitative than quantitative and closely related to Logic and its heir, Computing, probably the formal language which best captures the complexities of biological phenomena. However, any reader who is clued into the history of Biology will be familiar with François Jacob's *The Logic of Life*. This is a book of Natural History and Philosophy providing a perspective on Biology which is as well founded and coherent as could be written by one of the fathers of Molecular Biology. However, the approach I develop in the work does not attempt to use the term *Logic* in the same way that Jacob does, which coincides with the colloquial use of the term, namely to try and capture the common thread, the last reason which justifies something, in this case: life. The purpose of my essay is to reflect on a formal language which is suitable for biological description and explanation. I maintain that this language is close to Logic and Computation. This language could well serve as the Calculus needed in order to be able to theorise better about Biology, in the same way that Calculus has allowed with Physics. Hence: *The Calculus of Life*.

My first scientific–philosophical work was published in 1982. Telling the story which led me to it is of considerable interest when attempting to understand the central thesis and objective of this book. The work was an analysis and critique of Joseph Henry Woodger's axiomatisation of Biology (Moya 1982). It was the result of work on a thesis for my degree in Philosophy. Given my previous training and interest in Biology, my supervisor, Professor Manuel Garrido, who was then professor of Logic and Philosophy of Science at the University of Valencia, suggested that I should study the work of Woodger. In his opinion, the reason for doing so was blatantly obvious: Woodger was a biologist who had entered into the field of formalising scientific theories. There could be nothing better for a young biologist who also studied Philosophy than to devote his endeavour to such a project. I was excited about the proposal. Professor Garrido invited me to his house and led me to the section of his enormous library which was set aside for science. He showed me, amongst others, an essay by J.H. Woodger named *Biology and Language*. It was published in the emblematic collection of essays of Scientific Philosophy named *Structure and Function* (whose subtitle was *The Current Future of Science*) by Tecnos Publishing House and then run by the amiable Professor Enrique Tierno Galván.

Woodger aimed to achieve a level of formalisation in Biology which was on the same level as that which, in his opinion, existed in Physics. He believed that all sciences should meet the requirements of formalisation. Namely, using a series of postulates or initial axioms, it should be possible to deduce using the more or less arduous exercise of logical and/or mathematical reasoning, a series of contrastable predictions. In so much as such predictions were or were not empirically valid, the very initial axioms could or could not be suspected of being false. In essence, Woodger was claiming a hypothetical-deductive model for Biology which is a requisite of all mature sciences. In his opinion, such a structure would materialise much better if it possessed an appropriate formal language. This presents us with an initial question: what could this language be? This is not a trivial question as Mathematical Analysis and Calculus are probably not the most appropriate means, and we may need to use other more qualitative measures or those capable of reflecting the complexity of biological phenomena. Logic and Computing could be the most suitable languages.

Woodger was aware of the enormous difficulties posed by such an undertaking, as there are many elements to consider when talking about a living being and therefore about the process of formalisation. It is not simply that they have unique properties such as *autonomy*, *reproduction* and *evolution*, but rather that when we dissect live bodies, we find that they have the same properties at different levels of their organised hierarchy, as if they were fractal objects. Evidently, any formalisation ought, on the one hand, to consider the axioms which allow for the aforementioned properties to be deployed. However, it should also make considerations relating to how some levels give rise to others, as well as the interactions between them. Even thinking about this reveals the extraordinary complexity involved with such a task. Woodger was aware that formulating a general axiomatic theory of Biology was not an attainable possibility given the state of empirical knowledge about Biology at the time. However, this did not rule out the possibility of carrying out partial formalisations. Thus, looking into the aforementioned work, as well as Woodger's *The Axiomatic Method in Biology*, one finds attempts to formalise Mendelian theory, Embriology, Taxonomy or Evolution.

Woodger was proposing the need for a formal biological theory around the nineteen thirties. My undergraduate thesis was published 50 years later. I concluded my dissertation by saying that the climate was not ripe for such work and that we lacked fundamental knowledge of many processes and mechanisms operating in cells and at other levels of organisation of living organisms. A large number of Woodger's formal developments were made using definitions, and although at that time, it was possible to subsume or derive some of them by reformulating certain axioms, it is true to say that not much progress had been made in the attempt to formulate a sufficiently general theory. However, nor is it possible to say that advances in Biology up until the eighties of the past century made it possible to delve deeper into Woodger's dream of an axiomatic theory of that scope. This is in spite of the enormous developments made in Biology during the fifty years prior to the eighties, particularly in relation to the structure and role of DNA.

Woodger's work is an example of what could be viewed as a premature attempt to formulate a general theory of living beings. However, have there been other examples? How many previous attempts do we know of that were made by other scientists who have similar objectives and run parallel with or are even contemporaries of Woodger? There are several of course, but allow me to make reference to two who complement Woodger well in so much that they emphasise properties which are very relevant to living organisms, semantic properties, and contrast with Woodger's syntactic approach. I am talking about the pioneering works of Alan Turing and Robert Rosen. We can consider a living being to be a very special machine which is made up of two well-differentiated parts: on the one hand, there is the information program or *algorithm* and, on the other, the equipment or physical apparatus which carries this out. Although there are other precedents (such as the case of John von Neumann), I am interested in referring to Turing as he is one of the fathers of Computer Science and because Computing can probably be used as a very suitable language for Biology, in the same way that Calculus is with Physics. The other concept comes from Robert Rosen, who posits Biology as a *relational* science, which must concentrate on looking deeper at the concepts of autonomy and independence of the basic unit of life, the cell, in relation to the environment which surrounds it.

The syntactic–axiomatic approaches of Woodger and the semantic approaches of Turing and Rosen around the relation between the information program and its expression or the cell's autonomy, respectively, can be considered as fundamental and ground-breaking attempts to move us closer towards a theory of life. Although they are premature owing to a lack of detail, in other words, it would appear that detail is not simply a triviality in living things. Detail provides complexity, and we need to have a sufficiently realistic knowledge of it in order to make a robust attempt to formulate a sufficiently mature theory of life. It has taken another forty years, almost, for us to be able to begin to enter into the intricate complexities of the cell. I refer to the cell because although other levels exist in the organisational hierarchy of organisms, the fact is that it is fundamental to understand what happens in the cell in order to move beyond it and move into the realm of multicellular organisms. Only recently have we begun to witness attempts to create complete computational cell models. To do so, it is necessary to go beyond, and far beyond, the basic concepts of a theory of living things outlined by the illustrious predecessors, Woodger, Turing and Rosen, amongst others. It is convenient for us to say that we possess a formal axiomatic theory about living beings (Woodger) or that the cell is a machine which carries out an information program or algorithm (Turing). We can even say that the internal relations established between the different fundamental components of the cell are as important as the components themselves in order to delimit the fundamental possession of autonomy (Rosen). Although this is all necessary, it is insufficient for the simple reason that it creates an excessive abstraction of life's phenomena. The realism that we must introduce into theories about living organisms needs to go beyond simple models, minimal axiomatic systems or very basic outlines of cellular structure and function. It is probable that some recalcitrant theorist is inclined to say that modern Biology is simply about filling in the framework created by the above persons and other eminent forerunners. They may even say that the best way to understand complexity is to detect the simple patterns which create it. However, considering the extreme empiricism which surrounds contemporary biologists and biomedical researchers, what is certain is that that theory will acquire greater elegance in so far as we are able to recreate life rather than make it abstract. In other words, we must continue to move towards a realistic production of the great conceptualisations or abstractions of life, which would appear to be happening in current Biology.

In that regard, in order to create a computational model of a cell, we have had to build up a sufficiently detailed knowledge of its individual molecules and the way in which they interact over time. This has only recently been achieved with Genomics and other high-performance techniques which have enabled the characterisation of genetic and molecular composition of different organisms. Secondly, however, we cannot forget that the availability of all this information is insufficient as it is necessary to create models in order to gain a sufficient understanding of what takes place. Models form an important part of concepts referred to by Woodger, Turing and Rosen. Indeed, they are formal and axiomatic in so far as they are based on a series of assumptions, equations, restrictions, conditions of the surrounding area, etc. To a certain extent, and progressively more so, they are also computational, given that their behaviour is simulated on the computer and contrasts with the empirical results or the available measurements from real systems which have been studied. They also try to formulate corresponding models using the Rosen notion of autonomy and the relation between the different components of the simulated system. Indeed, many different models have been made over recent decades, each one focussing on some of the properties or particular components of the specific cellular function which it attempts to model. Progressively, models have looked at more complex and integrated functions. Considering that the computational model of a cell reproduces more faithfully the behaviours of the thing which it is trying to simulate, we can state that we are moving closer, with increasing realism, to providing empirical precision to the abstract ideas of primary theorists. I shall also have the opportunity to show that such abstractions have an important heuristic value—and still do—whilst we do not possess means to carry out appropriate empirical tests. To a large extent, we are dealing with mental experiments which have achieved great success in other sciences.

It may appear that this strategy has a final objective that we can aspire to something akin to a definitive theory about cells and by extension, about life, if we also include evolutionary theory. However, this is not necessarily the case. The dichotomy between chance or possibility and necessity is a fundamental stigma for life and evolution. The stigma is evidence of the presence of both, not simply of one or the other. There is an excessive insistence on emphasising the historical nature of life, when the other element of the dichotomy, necessity, is also implicit. What type of theory can we hope for, what type of predictions can be made, which have to make a judgment considering the active and effective presence of both acting forces? Throughout this essay, I will have the opportunity to demonstrate the relevance of results coming from other fields of knowledge, primarily Logic and Computing. The purpose of this was to answer the question about whether the evolution of biological complexity is necessary despite the continuous contingencies encountered along the way. I wish to point out, however, that this reference to necessity does not imply any finality or teleology. It is precisely the fact that life's evolution is affected by contingency which greatly humbles any attempt to devise definitive theories. Possible processes of generating increasing levels of complexity must be compatible with the fact that such levels are achieved by totally unexpected biological entities. Mammals evolved successfully and some of their descendants had a spectacular conception, resulting in beings with language and extraordinary levels of intelligence. However, it is probable that nobody would have been able to envisage such an event had they been able to examine the specific period when the planet was the realm of dinosaurs.

There are two reasons why I refer to the relevance of progress in Logic and Computing in relation to understanding the genesis of biological complexity. I hope these reasons will become gradually clearer. On the one hand, there is the notion that computing itself allows for a simulation of the origins and evolution of biological complexity. However, on the other hand, there have been advances in these scientific fields which limit the scope of what we can manage to explain or predict about life and its evolution.

The book consists of an introduction and twelve chapters grouped into three parts, which I have called: *Biology; Logic and Computing; and, The Cell and Evolution*, respectively. The reasoning behind the three names is as follows. In the first part, I concentrate on relevant precedents in Biology which have helped to shape current theoretical thinking about Biology. The second part aims to show that computing is a very suitable formal language for theorising about Biology. In the third part, I offer several thoughts which bring together the relation between Computing and Biology, particularly the cell and evolution, and the direction being taken by biological research on these matters.

The objective of the first part was to review the ideas put forward by some biological theorists that have had a considerable influence on shaping of modern Biology. I am not going to deny that the list of people chosen is biased due to the fact that they are authors who have influenced me personally, and people I would like to remember as pioneers. In Chap. 1, I emphasise the relevance of certain European thinkers in the field of the Philosophy of Life, thinkers of a certain continental style, so as to think about life in a way that does not confront scientific culture with the humanities. I have also used that chapter to show how evolutionary theory is a natural link between both cultures. I also offer some more specific considerations regarding the current state of research into evolutionary theory. Chapters 2–5 are special tributes to Monod (Chap. 2), Jacob (Chap. 3), Waddington (Chap. 4) and von Bertalanffy and Baquero (Chap. 5). I have only picked a small selection of concepts which, in my opinion, permeate many current considerations regarding theories about life.

The second part addresses computational models which attempt to capture properties of life. These are not of course the only ones in the recent history of the relationship between Computing and Biology. However, intuition has always led me to think that Logic and Computing were suitable languages for Biology. I look at this subject in Chap. 6. I will admit that such intuition may suffer from a lack of self-criticism, although the examples I look at in this chapter actually veer in the opposite direction rather than backing them up. This is the case with the program known as *Life* (Chap. 7), a cellular automaton created by John Horton Conway, which has a very interesting conceptual basis: a world of almost infinite possibilities which develops from a finite collection of rules. It is worth considering the relationship between this game and concepts such as *complexity, emergence, determinism* and *closed evolution*. Something similar occurs in the algorithmic Chemistry of Walter Fontana and Leo Buss (Chap. 8), where I present the fundamental developments of these authors which apply the concept of *recursion* and λ *(lambda)-calculus* in order to create structures which capture the basic properties of living organisms: reproduction, emergence and self-maintenance, to mention just a few.

The third part attempts to show the theoretical scope of current research concerning the cell and evolution. The practically unlimited universe of genotypes greatly determines the ability to succeed in posing high-level verifiable biological theories (Chap. 9). However, we cannot limit this potentially infinite universe exclusively to genotypes. If we seek to create a computational model of the cell, it is necessary to analyse not only the genetic memory, but also the additional epigenetic memories which lead us to final phenotypes (Chap. 10). Moreover, we cannot exclude inherent limitations which may impact on our ability to make predictions about the cell and evolution if we accept that both can be formulated in algorithmic terms (Chap. 11). Of particular interest are the recent theoretical findings which relate to the conditions used to make predictive theories. Finally, Chap. 12 offers a summary, and I seek to posit current Biology as the realisation of Goethe's dream and also refer to Kant's challenge that we would never be capable of producing a Newton in Biology.

Earlier versions of Chaps. 8 and 12 of this book have been published in cooperation with colleagues and friends. I would like to thank the many pleasant hours spent in related studies.

Chapter 11 has previously been published (Moya 2009) although I have rewritten and adapted it to accommodate the objectives of the present work. There is even a new section, specifically related to the possibility of a predictive evolutionary theory (Chaitin 2012; Day 2012). Chapter 12 has also been previously published (Moya et al. 2009) but adapted for the present work.

I wish to thank Antonio Moya for his invaluable help, his corrections and suggestions to improve the text, for the many hours he has devoted and his support. The fact he is my son did not affect the seriousness and thoroughness with which he undertook this task. I wish also to thank Fabiola Barraclough for her expert translation to English of the Spanish book and my assistant, Dr. Laura Domínguez, for taking care in preparing the book for this SpringerBriefs collection.

This work has benefited from EU funding (TARPOL, ST-FLOW and SYMBIOMICS projects), Spanish Ministry of Economy and Competitiveness (Projects SAF2009-13032-C02-01 and SAF2012-31187) and the Generalitat Valenciana (PROMETOII/ 2014/065), Spain.

# References

Chaitin G (2012) Proving Darwin. Making Biology mathematical. Pantheon Books, New York

Day T (2012) Computability, Gödel's incompleteness theorem, and an inherent limit on the predictability of evolution. J R Soc Interface 9:624–639

Moya A (1982) Panorama de la obra de J.H. Woodger en la Biología contemporánea. Teorema XX:61–70

Moya A (2009) Synthetic Biology, Gödel, and the blind watchmaker. Biol Theor 4:319–322

Moya A, Krasnogor N, Peretó J et al (2009) Goethe's dream: challenges and opportunities for synthetic Biology. EMBO Reports 10(Suppl 1):S28–S32

# Part I
# Biology

# Chapter 1
# Theoretical Biology and Evolutionary Theory

**Abstract** During the nineteen-sixties and seventies, several biologists made decisive contributions to the development of evolutionary thought and theoretical biology, smoothly bridging the gap between science and humanities, and overcoming the classic antagonism Snow claimed to exist between the two worlds. Scientists like Jacob, Lorenz, Monod, Rensch, von Bertalanffy or Waddington, to whom I pay homage here, are a select group of scientists pertaining to the so-called third culture, who have staged a scientific assault on Natural Philosophy—the forerunners of subsequent science popularisers such as Dawkins, Gould, Penrose, Gell-Mann or Pinker. Finally I call for a more detailed study of the recent history of Theoretical Biology, to put it on a par with that undertaken for Evolutionary Biology, with an understanding that the latter is a pillar of the former.

In my work *Evolution: a bridge between the two cultures* (Moya 2010) I posit that biological evolution is a place where scientific and humanistic cultures meet naturally and that the way to overcome the antagonism between the two is through multidisciplinary study, something already disclosed by Snow in his famous book (Snow 1977). I criticise what is known as the *Third Culture* mentioned in Brockman's thesis (1996), as well as the fact that the only references on the scientific assault of Natural Philosophy were exclusively centred on Anglo-Saxon authors; mentioning Dawkins, Gould, Penrose, Gell-Mann and Pinker as a few examples of well-known scientists. I stated that there is a continental European culture in which these two traditions fuse, in issue that should probably be addressed in depth and is not necessarily reflected in the contributions made by the authors mentioned above. What is more, they were forerunners of the above writers. In fact, I do not consider that the position of continental authors in the first half of the twentieth century, such as the German (Rensch 1971; Lorenz 1972; von Bertalanffy 1976) and French (Monod 1972; Jacob 1973) scientists and philosophers represented a scientific assault on the humanities. Rather, I see it more as an integrative and synthetic Natural Philosophy strictly speaking. In other words: to what extent are they not the best exponents of the third culture, and even historically ahead of the others mentioned by Brockman? In my 2010 book I argue that the proposal

© The Author(s) 2015
A. Moya, *The Calculus of Life*, SpringerBriefs in Biology,
DOI 10.1007/978-3-319-16970-5_1

by these continental authors is genuinely inclusive: Biology and biological evolution facilitate the encounter between physical sciences, life sciences, social
sciences and humanities. In this essay I am going to recount in some detail the relevance of some of the aforementioned authors, specifically, Monod, Jacob and von
Bertalanffy. But I shall also defend a heterodox British author who, in my opinion,
has been crucial to defining theoretical biology: Waddington.

## 1.1 Four Great Biologists

Between 1972 and 1976, four works were published in Spanish which I consider to
be fundamental because they can be seen as anticipatory in the field of natural philosophy on life and theoretical biology. They are *Chance and Necessity: Essay on
the Natural Philosophy of Modern Biology* (1972) by Jacques Monod; *The logic
of Life* (1973) by François Jacob; *General Systems Theory* (1976) by Ludwig von
Bertanlanffy; and *Towards a Theoretical Biology* (1976), edited by Conrad Hal
Waddington. These works have had a special influence on both my personal and intellectual development. On a personal level they fuelled my desire to know about other
worlds and formations. Intellectually, they unfolded the challenge of the complexity
of living things and exhibited natural philosophy. Although this philosophy is com-
mitted to science, it is no less critical of certain reductionist trends in the domain,
which could be harmful to the theoretical development of understanding what life is.
It is worth reflecting on them because, over time, I have come to realise that my own
thinking has developed as an extension of those readings, which I have revisited regu-
larly. Although I might not have been particularly aware of it at the time, I have realised of late, that they have played a decisive role in my particular way of conceiving
and addressing research into life. Interestingly enough, none of them have as the central object of study the theory of evolution or philosophical reflection on it, although
logically, all of them have it present or mention it in some way. I wish to explicitly
mention this fact because evolutionary theory is my area of expertise, and it may
seem contradictory that I have conducted my research steps towards the development
of evolutionary theory and not to other areas of research closer to the ideas pondered
in the works above. Much of all modern Biology is a reconfirmation of the common
origin of all living things. Such an origin also entails the confirmation of common
processes and mechanisms underlying the functioning of all beings or, in any case,
our ability to determine at exactly what point they appeared in evolutionary history.
This essay is an attempt to show the solution to this contradiction. Against the back-
ground of biological evolution, this essay aims to offer and kindle reflection on the
tremendous problems posed by understanding and explaining biological complexity
in all its dimensions.

Besides reflecting on this continental approach (with the exception of Waddington)
to Natural Philosophy and Theoretical Biology of living things, I think it is also
appropriate to refer to other great thinkers from different geographical origins and
fields of research that played an important role in the development of theoretical

thinking in Biology during the second half of the past century. I refer to Alan Turing and Robert Rosen. Both had, in turn, important forerunners, who we would have to place in the first half of the last century, particularly Gödel. Both this mathematical logician and Turing are relevant for different reasons, the first being that their theorems have permeated the empirical sciences. By assuming, somewhat riskily, that the laws of nature are algorithmic, it can indeed be derived that unpredictable behaviours may occur in nature from the set of axioms that characterise them. Throughout this essay I will refer to Gödel on various occasions. Turing is fundamental to the extent that he contributed to the development of Computing Science, which represents a highly appropriate language to capture the complications of biological phenomenology. Of these three aforementioned authors, Rosen was probably the keenest to understand what life is, as he tried to orchestrate all theorisation around this question, particularly the cell, capturing its abstract properties (isolation of the environment, metabolism, self-maintenance) and formulating a frankly pioneering Theoretical Biology based on the relational properties of biological entities.

## 1.2   Theory of Evolution and Compositional Evolution

Why have I not considered the theorists of Evolutionary Biology in depth in this essay? Fundamentally I would say that the contributions made to Theoretical Biology have been very well covered by the theorists who fathered Evolutionary Biology, namely Fisher, Haldane and Wright. From their contributions, there has probably developed a certain train of theoretical thought that has neglected the contributions of the other authors given greater prominence herein. Even so, there should be no doubt that Fisher, Haldane and Wright are always present. Their formulations are a kind of null hypothesis on which the evolutionary theory itself has grown.

Dobzhansky's dictum is well known, he is considered the empirical successor of the previous three, stating that "nothing in Biology makes sense except in light of evolution." And indeed, we can hardly think how living things function and how they are organised if we do not consider how their functions and organisation came about. Evolution introduces an element of time and history that cannot be ignored. Another question is whether the explanatory theory derived from the above theorists about the emergence of biological organisation and functions is enough. Dobzhansky's expression does not say that we have to resort to a particular theory of evolution; nonetheless, he does emphasize that evolution is absolutely indispensable. This is because in any theory of biological organisation we need to incorporate the presence of factors that contribute in evolutionary time to the creation of a heritable genetic variation. This variation is not exclusively limited to an opportune mutation. There are other important sources of variation, and these are the same sources which modern Evolutionary Biology is revealing, thus complementing the theory of gradual evolution. Although many theoretical and experimental studies have been published in recent decades with the gradual expansion of the theory of evolution, we find an excellent example of synthesis and conceptualisation in

Watson (2006). The reason why I am greatly interested in introducing this author's work is because many of his contributions to the non-gradual evolutionary theory are not computational, which reaffirms the value of this science, its language, methods and concepts in helping to consolidate theoretical biology.

Watson defines a type of evolution that is not compatible with gradual evolution, which he calls *compositional evolution*. The author handles this issue of non-compatibility with care, endeavouring to anchor it in a well-known biological phenomenology by giving empirical support to construct conceptualisations and simulations that show us these phenomena result in non-gradual processes. Under the compositional heading, Watson groups the evolutionary processes that involve combining biological systems or subsystems of preadapted genetic material. The concept of *preadapted* must simply be regarded as systems or subsystems that are adapted prior to being integrated. The typical phenomena to be included in this notion of compositional evolution are phenomena such as sex, hybridisation, horizontal transfer and allopolyploidy, but especially phenomena as such as genetic integration of symbionts, or any other mechanism that encapsulates a group of simple entities into a more complex one, providing a new level of organisation. One can see that this evolutionary concept is introduced with empirical grounding. Symbiosis, for example, is a field of research that has taken an unprecedented place in modern Biology and it is increasingly common to find examples of integration like those mentioned by Watson (Moya and Peretó 2011). Moreover, compositional evolution could go beyond the simple establishment of symbiotic systems. Consider, for example, the extraordinary complexity of the associations that exist between certain organisations of typical eukaryotic and microbial communities, usually bacteria, which are prokaryotes. These communities are known as *microbiota* and it does not take much to understand that multiple interactions of all kinds can take place between all those involved. Current research is avidly investigating how such associations have evolved on the phylogenetic scale, and the process and nature of the ecological succession established during the ontogeny of host organisms. Furthermore, we cannot fail to see that such evolution (which could be called *cooperative*) entails or should entail advantages for those involved and, therefore, the evolution of new entities must in turn respond to Darwinian selection.

I have already mentioned Waddington as one of the pioneers of Theoretical Biology, and I will to do so again later. Indeed it was Waddington who formulated the notion of the *epigenetic landscape* to show that the phenotype of an organism is the product of complex interactions between genes and many other factors present throughout development, among which we should not rule out accidents or randomness. Certain combinations of these factors are channelled and lead to particular phenotypes. If we take this concept of an epigenetic landscape and add the microbial component as another factor, we see that the resulting phenotype is the combination of multiple factors involved, entailing not only the genotype of the host but also the microorganisms it harbours. The latter are subject to restrictions and boundary conditions, and are in constant interaction with their host and environment. That concerning the host genotype can also be applied to the microbial community it harbours (e.g., general properties for the generation of genetic variation, different levels of gene expression regulation, metabolism, functionality of the system when faced with disturbance), but we must also bear in mind that these are new genetic entities, which become part of the whole and must, therefore, be at higher hierarchical levels to somehow ensure proper dialogue between them.

As Watson did, I believe it is worth pointing out the difference between compositional evolution and *saltational evolution*, because we could get the impression they are similar when in fact they are not. The supposed genetic changes occurring in evolution are saltational as a result of an increased rate in the accumulation of small changes, which is often very high in short intervals of geologic time. On the other hand, we could also think of large-scale mutational processes, including their effects on the corresponding phenotypes. However this would not be considered compositional evolution, because such changes occur in the genetic makeup of certain bodies. The same could be argued of phenomena such as gene duplication. These processes are not able to substantially increase the complexity of the organism undergoing them. They are phenomena that occur in the genomes of adapted organisms, and their fate is dictated by natural selection. For Watson, Darwin's gradual evolution incorporates both classic gradualism and saltationism, because both forms of evolution are essentially linear. This means that the accumulation of random genetic changes is linear, whatever their size, although probably the most effective changes are small in magnitude, since their phenotypic effects have to be evaluated by natural selection. Compositional evolution speculates on the value that the combination of pre-adapted genetic modules may have for evolution in general, and examines how the basis of such a circumstance could generate evolutionary novelty or complexity. The strategy of this success lies in the fact that the merging genetic components have previously adapted. Moreover, on the basis that each one is probably involved in different functions, they may well lead to a successfully evolved new entity, responding effectively to natural selection. While the model of gradual evolution works by a successive linear approximation to an optimal point, the compositional model practices a different strategy: to divide and rule. The integrated systems or modules are functionally pre-adapted to perform different functions; and if integration is successful, it may be more efficient than the separate systems.

An important final point: Let us look at how the concept of evolution by natural selection is present both in gradual and compositional evolution. When I have referred to gradual evolution as a form of linear evolution, saltational evolution was also indicated as such. We should ask whether another form of evolution, the so-called *neutral evolution* can be seen in the same light as gradual evolution, i.e., linear-like evolution. Personally, I do not see this issue as clearly as Watson. It is true that neutral evolution is not controlled by natural selection, since it is not able to discriminate between selectively similar variants. Therefore, its evolution does not depend as much on natural selection as it does on population size, known as genetic drift. Therefore, the accumulation over time of many neutral genetic changes could, at any given moment and in new environmental conditions, enable radically new phenotypes to emerge. But for Watson this would amount to a major genetic change whose evolutionary success would probably be frustrated on confronting natural selection. Therefore, the nature of neutral changes and the rate at which they occur would remain linear.

# References

Jacob F (1973) La lógica de lo viviente. Editorial Laia, Barcelona

Lorenz K (1972) El comportamiento animal y humano. Plaza y Janés, Barcelona

Monod J (1972) El azar y la necesidad. Barral Editores, Barcelona

Moya A (2010) Evolución: el puente entre las dos culturas. Editorial Laetoli, Pamplona

Moya A, Peretó J (2011) Simbiosis. Seres que evolucionan juntos, Editorial Síntesis, Madrid

Rensch B (1971) Biophilosophy. Columbia University Press, New York

Snow CP (1977) Las dos culturas y un segundo enfoque. Alianza Editorial, Madrid

von Bertanlanffy L (1976) Teoría general de sistemas. Fondo de Cultura Económica, México

Watson RA (2006) Compositional evolution. The impact of sex, symbiosis and modularity on the gradualistic framework of evolution. MIT Press, Cambridge

# Chapter 2
# Chance and Necessity

**Abstract** In *Chance and Necessity* Monod masterfully spans the divide between humanities and sciences. For Monod, unlike subsequent renowned popularisers, *science* was one thing but what science *evoked* was quite another; he claimed that we should take the logic of science as far as it allowed. In this chapter, I revisit Monod's reflections on Bergson—as did the famous evolutionary biologist Mayr—and I suggest that this philosopher's vitalist theory should be reconsidered in the light of modern Biology, as should Driesch's Embryology. The thesis on the need to give similar importance to both the parts of a biological entity as to the interactions between them is discussed within a wider context.

As soon as I could lay my hands on it, I read *Chance and Necessity* (Monod 1972) with fervour. The fourth edition, published in Spanish by Seix Barral, is a gem that I frequently reread in small snippets, intending to digest it slowly. The work is subtitled *An Essay on the Natural Philosophy of Modern Biology*. I wonder how many scientists of my generation and beyond have been influenced by its enormous intellectual scope. I honestly cannot answer that, although I fear that it is far fewer than it should be. Monod was perfectly aware that his book did not respond to a systematic treatise of Philosophy of Science nor a formal work around the crucial experiments that led to the birth of Molecular Biology. Monod's eagerness, like many other continental scientists, was to do with transcending reflections on his own research to include the layman's thoughts about life. Monod says:

> Indeed we should avoid confusing the ideas that science suggests with science itself; but we must take the conclusions that science draws to the limit if we are to reveal their full significance.

Monod clearly airs something that subsequent scientists have not envisaged with such precision or demarcation in similar written essays: the distinction between what science *is* and what science *evokes*. However he is also in favour of taking the evocation of science as far as possible. This is not a worthless intellectual exercise. Science has the same right as other forms of thinking to interpret the world and to go as far as its rationality allows. This exercise is not futile as

© The Author(s) 2015
A. Moya, *The Calculus of Life*, SpringerBriefs in Biology,
DOI 10.1007/978-3-319-16970-5_2

it allows free competition with other systems of thought—especially Philosophy, Religion and Art—to propose consistent explanations for the world. Nevertheless, this is neither a dominant nor domineering way of thinking because after a new scientific discovery, old problems may acquire a new interpretive light, however naive or harmless this finding might be. One must be there to transcend the discovery, and help place it in the context of modern culture. Monod assumed this task, aware that he could be treated as naive by philosophers and with distrust by scientists.

For Monod, vitalism and its scientific explanation of life is not an outdated, dark or oblivious philosophy. Monod tries to understand the underlying message in Bergson's *Creative Evolution*, which he labels as metaphysical vitalism. All the same he acknowledges that there is something intriguing in the French philosopher's *vital impetus* or *vital force*—the engine of evolution. This force is responsible for life's ability to tame and to organise the inanimate. We might think such a philosophy to be finalistic, with opportune determination for final and efficient causes leading to sustain a relationship between creationism and metaphysical vitalism. Not at all. We may wonder in hindsight as to the nature of that vital force, as well as trying to understand whether Bergson would have been satisfied by the current theoretical and empirical knowledge about the cell and its functioning, thus being perfectly integrated in science. I like to imagine that it would be so, for this thesis goes in favour of bridging the gap between philosophical and scientific rationality. With the insufficient conceptual and empirical tools of his time, Bergson was unable to give a scientific answer to that question. But he knew that there must be something fundamental in life leading to unfold states or processes that were unique with respect to other things in the world. In the absence of such an explanation, the philosopher grasped at an intangible principle. As a working hypothesis it was not bad, because above all there was a need to find some explanation of its uniqueness, particularly related to the domination and control of the inanimate. Many times we have resorted to the formulation of entities without immediate or available physical correlates, which have become the basis for an explanation of world phenomena: force, inertia, genes, the interaction between components, to name but a few.

In evolution, Bergson recognises that man represents a supreme state; however he does so under total indeterminacy. We are a wonderful manifestation of the creative force, but just as we are here, we might well not be. We may well deliberate over the origin of this vital force just as science ponders on the origin of life. Once life appears, so does this force. In fact, they would be equivalent. To provide a theory that explains the origin of life would also explain the origin of this force. But such as theory was not available in his time, and indeed today we are still engaged in this endeavour. According to Bergson, evolution has led to man in an indeterminate way, and man is an entity that demonstrates the total freedom of this creative force. Such metaphysics is not incompatible with science or the explanation of the origin and evolution of life.

Man is most certainly a product of evolution. For Bergson, however, our rational intelligence is something that has allowed us to dominate nature. But this quality is not enough to grasp what life is. According to this philosopher, such apprehension can occur by resorting to another general quality which the vital force bestows on every living being: instinct. Instinct can endow man with the intuition needed to capture life in all its dimensions. This thesis cannot necessarily be shared as it is a retrospective view glimpsed from current science. Because human intelligence, as we know it, is more than intellect adapted to dominating things or controlling the world. Intelligence is more than rationality. Intuition is one of myriad manifestations of intelligence. And man, in all the uniqueness implicit in deploying this intelligence, is a being that can look back on the path that has generated him. We know more or less where we stand on the sinuous tree of life and at which point this branch appeared. We are quickly approaching the point at which life will become intelligible, a point Bergson believed could not be reached by rationality alone. My question, again, is whether he would maintain this opinion on learning of all the new advances that have occurred during the twentieth century around the cell and the evolution of organisms.

And what of Hans A.E. Driesch's *entelechy* or vital force? For this scientist, who has also been called a vitalist, the cell is not captured in its components. When we delve into the innermost elements, we will discover components or processes that clearly assimilate physical and chemical phenomena. But it is not possible to understand what the cell is, or describe it by such material components. For Driesch it must hold an entity which he qualifies as non-spatial, intensive and qualitative and he had to resort to this explanation to state that a cell, the smallest unit of life, differs from an inert physical entity. It is true that the cell is composed of extensive and quantitative spatial entities, but their mere combination cannot capture the actual essence of the cell. There is something else, a property that endows its essence: the quality of life. Driesch, like Bergson, was far from catching a glimpse of such an entity, which had the remarkable ability, unique in the world, to evolve and modify the beings it imbued.

In 2002, during the Walter Arndt conference on the autonomy of biology, in a fey attempt to justify the issues raised by the vitalist authors, Ernst Mayr said that:

> It would be ahistorical to ridicule vitalists. When one reads the writings of one of the leading vitalists like Driesch one is forced to agree with him that many of the basic problems of Biology simply cannot be solved by a Philosophy as that of Descartes, in which the organism is simply considered a machine. The logic of the critique of the vitalists was impeccable. But all their efforts to find a scientific answer to all the so-called vitalistic phenomena were failures. Rejecting the philosophy of reductionism is not an attack on analysis. No complex system can be understood except through careful analysis. However the interactions of the components must be considered as much as the properties of the isolated components.

Mayr could not have given greater credit to the vitalists. To begin with he claims that they were not willing to assume that a living being, a cell in its simplest expression, was a machine. The Cartesian conception equating a machine with a live being is rejected. According to Mayr, this equivalence is the cornerstone of a reductionist philosophy that vitalist scientists like Driesch were unwilling to accept. Even in retrospect, we can now clearly state that if a living being were like a machine, it would be a very special one indeed. It has a particular ability to overcome problems and defects and still function despite them, as long as they are not overly destructive. But unfortunately vitalist scientists had a monumental problem: even with the conceptual tools, techniques and empirical data of their day, they did not find a convincing scientific explanation of what a living entity might be. The value that the analytical study of living things to shed light on the intrinsic complexity associated with them was not denied. Indeed, Mayr weighs up the value of the analysis in the right measure. However akin to the former vitalists, he believes that life should also be studied from other approaches. Mayr makes his claim about the vitalists because they are forerunners of a thesis he finds attractive: Biology is an autonomous science. Furthermore, this scientist tries to put this autonomy into the current context. And that is when he introduces the very modern notion of *interaction* between components. For Mayr interactions are as important as the components themselves. Therefore, the following question should be framed: do the interactions of the components constitute a key element of the whole, one that is essential to give a comprehensive explanation of that entity we call life? Is the interaction, in a generic and qualitative sense, the vital force so keenly sought by Bergson or Driesch's entelechy? Most probably. This work is an intellectual journey guided by the notion that interaction is characteristic of the complex phenomena occurring in nature, associated to life and, particularly, to the cell, its basic unit. Emergent entities arise from interaction, and these can provide new levels of organisation that obey their own logic and their own laws. Life is an emergent phenomenon and represents a level of organisation with its own laws, which are required to understand the elements that compose it. The historical weight given to this issue by holistic biologists and philosophers is such that we could almost say it has been a secular tradition, running in parallel to others of a more analytical nature. An ever present tradition, critical and in minority, but unrelenting. A tradition that has been waiting for its moment of glory. Several factors have had to materialise to be able to affirm that the time for biological holism has come. Biology, or some of its traditions which are little inclined to analysis, has been a flagship for emergentism, upholding the thesis that each level of organisation has its own laws. Indeed, biology has instilled such a philosophy in other sciences, including Physics.

# Reference

Monod J (1972) El azar y la necesidad. Barral Editores, Barcelona

# Chapter 3
# Tinkering and Evolution

**Abstract** Jacob is one of the fathers of modern Biology, and in his monumental work *The Logic of Life* he recounts the polarity that has always existed in biological science between analytical and synthetic traditions, between reductionism and holism. Interestingly, however, although he is a model example of the reductionist tradition and one of the fathers of Molecular Biology, Jacob does not rule out the possibility of emergent properties.

As soon as the Spanish edition came out, I read *The Logic of Life* by François Jacob with passion (Jacob 1973). One of the reasons why I consider it to be one of the intellectual treasures in my library is because the author wrote me a dedication, coinciding with his designation as *Doctor Honoris Causa* by the University of Valencia in 1993. I clearly remember his eyes brimming with emotion as teachers and researchers, some already getting on in years, approached him the day of the ceremony with well-worn copies of his 1973 edition to sign. His work entered our country the best possible way: through the youth from my generation, eager to have a reference figure representing new Biology. Unlike Monod's more philosophical work, Jacob's work is an overview of the history of modern Biology. Even so, his review is interpreted from a scientific position which he calls *thomistic* or *reductionist Biology*. Jacob says that, contrary to what is frequently thought, Biology is not a united science. The heterogeneity of its objectives, the divergence of its interests, the variety of techniques, all help to multiply its myriad disciplines. There are two major trends at either extreme of this wide-ranging field, two views that are radically opposed. The first of these can be described as integrative or evolutionary. The integrative biologist refuses to believe that all the properties of a living being, its behaviour, its achievements, can be explained in terms of molecular structures. For this biologist, Biology cannot be reduced to Physics or Chemistry, but neither does he wish to cite the inexplicable vital force. Rather he believes that integration at any level, bestows upon the system properties that its elements do not possess singly. The whole is not just the sum of its parts.

At the other extreme of biological thinking, we find the so-called Thomist or reductionist approaches. Thomists believe the organism is a whole that can be

© The Author(s) 2015
A. Moya, *The Calculus of Life*, SpringerBriefs in Biology,
DOI 10.1007/978-3-319-16970-5_3

explained by the properties of its parts. Thomist Biology seeks to explain the functions only through the structures. Sensitive to the units of composition and functioning observed in the diversity of living things, it seeks the expression of its chemical reactions in the organism's achievements. According to this view there is no expression of the organism that cannot be described in terms of molecules and their interactions. It does not deny the integration and emergence phenomena, as the whole may have properties that the constituent parts lack. But these properties result from the structure of these constituents and their disposition.

Jacob argues that Biology is not a united science because—given its enormous observable and methodological heterogeneity—there are two extreme stances that directly oppose each other. The question is whether a general theory should bring these extremes together and unite Biology as a science. That is probably the goal of a theory of Biology: Biology with a theory. Jacob recognises that Biology rests upon many generalisations and little theory. Evolution is one of the few theories to enjoy this privileged position, albeit facing certain predicament from both inside and outside Biology itself. Jacob recognises that historicity is a problem for this science, as it is based on the reconstruction of events with little direct verification. Both the above and the previous considerations clearly show how Jacob came to be placed within the domain known as modern biology, once the structure of DNA was discovered. Indeed, Biology has since become a powerful discipline given its ability to delve into the very depths of the cell, the structure and function of molecules, and the interactions between them.

If, on the one hand, Mayr decidedly advocates the autonomy of Biology, declaring that living things exhibit properties at some time or level of their organisation or hierarchy that are not reducible to their components, then Mayr would be a good example of integrative Biology. While, on the other hand, Jacob is inclined to view that the components, essentially the DNA, the program, holds all the properties; including the emerging ones. Thus Jacob is closer to Thomistic or reductionist Biology.

It is important not to lose sight of something important Jacob realised: he might favour reductionist biology, but he admits emergent properties and interactions. Molecular Biology in its genesis is probably very much geared to the practice of experimental methodology, which tries to isolate problems and unravel the structure and function of the individual components. It is another question, however, whether these are the molecular components, and if in their very structure they hold the ability to generate emergent properties on interacting together almost mechanically. Even so, in my view, it is important to state that emergence is contemplated as something admissible by Jacob, an important biologist in modern Biology of reductionist bent. Jacob is in fact immersed in this historical trend of Biology that contemplates the vital phenomenon. In fact, in his book, he cites Claude Bernard's statement about what is living:

> Even admitting that vital phenomena are expressed physico-chemically, which is true, the question as a whole is not clarified; because it is not a chance encounter between physico-chemical phenomena that constructs a living being following a plan and a

pre-established or fixed design.... Physico-chemical conditions of vital phenomena are strictly established; but at the same time they are also subordinated and occur as a chain reaction, according to a previously established law: eternally repeated, in a certain order, with regularity, consistency and harmony so as to achieve the organisation and growth of the individual; be it animal or vegetable.

Claude Bernard also recognises that the way in which living beings are ordered and operate follows a particular pattern. He is able to do so from his experience as a father of modern physiology, and without having to refer (as did Bergson and Driesch) to entelechy or any kind of vital force. Such order emerges, is directed or controlled, but never by something external. It has an internal order, coming from within the living being. The physical and chemical processes that occur in it are not random but fully directed.

How far can we take this polarisation of Biology? The study of living things is complicated because of the extraordinarily complex and diverse phenomenology classified under *life*. Even so, just as Jacob stated, the truth is that there is an ongoing and unresolved tradition of integrative-evolutionary and Thomist-reductionist schools of thought. As I said, this gap could be bridged by the genesis of a theory of life, and a Theoretical Biology that admits reflections upon the Natural Philosophy of living beings. In the third part of this essay, I will outline and develop these two conceptions of Biology, which I will respectively call *synthetic and analytical*. There is not a complete equivalence between my notion of synthetic Biology and integrative-evolutionary Biology, nor between analytical Biology and reductionist Biology. The main reason is that reductionism admits multiple definitions that are well developed in the Philosophy of Science, particularly in Biology, to which Jacob is assigned. He may well represent a mixture of two of them: an extreme one represented by *ontological reductionism* and another less familiar one known as *methodological reductionism*. Ontological reductionism is equivalent to Jacob's conception that biological molecules, particularly DNA, are like a computer program that contains everything required to understand all forms of life, including emerging forms. Methodological reductionism is typical of the experimental sciences, or one might say science in general. Indeed, to study any complex phenomenon one must scrutinise its component parts.

It is important to delve deeper into the concept of emergent properties. As I shall have occasion to show in the second section of this work, I believe that the computational approaches endeavouring to simulate the patterns underpinning life may shed light on whether what emerges in evolution has, or not, autonomy from its component parts. In other words, it is important not to take reductionism as a starting position, but rather have hard evidence, in the form of calculation, experimentation or data, of the existence of emergent properties.

# Reference

Jacob F (1973) La lógica de lo viviente. Editorial Laia, Barcelona

# Chapter 4
# Concepts for a Theoretical Biology

**Abstract** Waddington is the father of Theoretical Biology and many concepts that are now the common currency of Biology-based language were coined by him, such as channelling, robustness or epigenetics. Waddington believes it is essential to work on a theory of the phenotype. The deployment of genotype to give rise to phenotype brings into play many of the laws required to formulate a theory of life. In recent times, Waddington is the best example of an advocate of cell theory; however, he was not in favour of a gene-centred approach to the study of biological evolution.

In the preface to *Towards a Theoretical Biology* Waddington (1976) states that:

> Theoretical Physics is a well-recognised discipline, and there are Departments and Professorships devoted to the subject in many Universities. Moreover, it is widely accepted that our theories of the nature of the physical universe have profound consequences for problems of general Philosophy. In strong contrast to this situation, Theoretical Biology can hardly be said to exist as yet as an academic discipline. There is even little agreement as to what topics it should deal with or in what matter it should proceed; and it is seldom indeed that philosophers feel themselves called upon to notice the relevance of such biological topics as evolution or perception to their traditional problems.

The first volume of this work was published in English back in 1969. So what has happened since then within the panorama of Biology to enable us to now state that we are approaching the prospect of developing a Theoretical Biology? Something truly amazing and which, in my opinion, puts us in a better position than ever to formulate a Theoretical Biology based primarily on the cell, its origin, structure and evolution. One main objective of this essay is to discuss this issue in depth. But first let me tell a little anecdote that may help to illustrate this idea. A few years ago I had the chance to visit the famous Institute for Advanced Study in Princeton, where Einstein and Gödel worked, among many others; a centre devoted exclusively to theory in many different fields of thought. I was greatly heartened to see that this centre had recently joined a programme in Systems Biology. The director, and other members of the programme, told me that Biology was mature enough to enter the Institute through the front door. The reason for

© The Author(s) 2015
A. Moya, *The Calculus of Life*, SpringerBriefs in Biology,
DOI 10.1007/978-3-319-16970-5_4

this lay, first, in the current availability of massive amounts of data on the cell, which until recently was considered a black box, and on multicellular organisms, populations and ecosystems; and secondly, the possibility of formulating thought experiments, in Einstein's traditional way, and being able to put them to suitable empirical tests using new genomic technologies and modern computational techniques that allow the simulation of life.

The introductory chapter of Waddington's *Towards a Theoretical Biology*, a work devoted to reflecting on the problems inherent to establishing a Theoretical Biology at that time, is still an authentic inspiration for thought on the elements around which Theoretical Biology should revolve today. Indeed the structure of the sections and the content covered in the chapter entitled *The basic ideas of Biology*, largely anticipate what may well be the future of the discipline, with little variation. What has happened since the nineteen-sixties merely provides empirical support, realisation and materialisation of the conceptual programme Waddington outlined in this celebrated chapter. Waddington probably represents the prototype of a theoretical biologist and that chapter, as mentioned above, could serve as a framework, referred to retrospectively from current Biology, on which to construct a Theoretical Biology.

The core of Waddington's proposal is what he calls *the theory of the phenotype*. The basic argument is, among other things, against limiting conceptions of what the living organism could be. First he ponders the relative importance of the genotype compared to the phenotype in defining life. The genotype is a component of the cell, but the extent to which it harbours specificity, and such specificity is transmitted to offspring, has become a not too well justified precedence over any other part of the same. Genocentrism has been relevant to consolidating Genetics, and has probably maintained its status with the advent of Genomics and its application to the study of evolution and reconstruction of the tree of life. We have gone from the gene to the genome and from genocentrism to *genomecentrism*. The close relationships between species can be established without the need to go into their physiologies, and this could give the impression that we do not need to know how that physiology unfolds in particular organisms in as much as each one transmits its specificity (information) to its descendants, both the immediate and more distant ones, in genealogy. True to say, the specificity associated to genomes changes over time by the action of evolutionary forces, such as mutation or any other factor that alters the size and composition of the genome, be it natural selection, genetic drift, and so on. Indeed, the comparative study of genomes is a very active field nowadays. Authors such as Koonin (2011) dare to support the existence of certain universal principles for evolution, as follows:

> Many, if not most, gross patterns of genome and molecular phenome evolution are shaped by stochastic processes that are underpinned by the Error-Prone Replication principle and constrained by purifying selection that maintains the existing overall (but not specific) genome architecture and cellular organisation.

Here is not the place to discuss the full scope of all that underlies this sentence, which is of importance. But it should be pointed out that genomic study through evolution, as has occurred so far and culminating in this quotation by Koonin, makes a clear abstraction of what an organism could be or, in any case, its smallest expression: the cell. Koonin's molecular phenome refers to proteins and their proper folding within the intracellular concoction. Koonin believes that both the genome and proteins taken individually are subject to stochastic laws corresponding to dynamics, based on the error inherent to the specificity principle—genetic matter—and the elimination of those variants that are not sufficiently effective by purifying selection. The panoply of genomic architectures that appear here and there in populations of different organisms, and the protein structures associated with a particular cell structure, correspond to averages that simply follow these stochastic laws, which Koonin termed as the *universals of evolution*. Not a single reference is made to the functional unit, which may be the cell, or the physiology, which could be the organism. The cells or organisms are theoretically unattainable products, because they behave like gases in statistical mechanics, with the genes in the genomes being equivalent to the gas particles. Moreover, and as a corollary, he dares to state that the importance of evolution is not so much in what happened once the cell has formed (the rest is history, says Koonin), but in changes in the replicons and proteins prior to the formation thereof. In any case, if we welcome the fact that universals are found in the evolution of genomes, an issue I believe relevant, the important question is whether the cell, or even the cell in evolution, is more than its genome, or its genome in evolution.

For Waddington, the information theory model put forward by Shannon and Weaver did not capture the complexity involved in an organism's development starting from zygote formation. The said information theory is a model to determine how the amount of information changes from an original source A to a receptor C by virtue of being transmitted through a channel B; and how, depending on whether or not this channel is closed to receiving information other than that coming from A, the information reaching receptor C whether or not quantitatively different to that from A. It is true, however, that this conceptual framework can abstract a certain variety of biological phenomena, such as the transmission of nerve impulses or even the transmission of genetic information from one organism to its offspring, particularly when organisms are simple in their genetic makeup. But the complexity of what happens in channel B is that the cell distorts greatly and efficiently the information passing from source A to receptor C. Waddington called for a theory to understand how this genetic information is displayed throughout the time axis of the developing organism: a theory of the phenotype. It is remarkable to realise that he was the first one to use the terms *epigenetics* and *canalisation*, buffering or robustness of the phenotype. Rather than a set of unrelated traits controlled by single genes, it is the phenotype that is considered an epigenetic phenomenon, as the "synthesis resulting from a complex interacting system, the total epigenotype". According to Waddington, epigenetics would study the interaction between genes and their products, interactions that would confer the phenotype on the being. With respect to robustness, Waddington clearly states the need to find explanations for the ability of the phenotype to restore a certain pattern despite external disturbances, or the fact that certain alterations in the genotype do not necessarily lead to appreciable changes in phenotype formation, as if the phenotype was buffered against such effects. He also entertained the idea, which could be important in Theoretical Biology, that the cell could be considered as an oscillator to the extent that it often exhibits negative feedback phenomena that could be the basis of oscillatory behaviour. So, what could one expect from the interaction of intracellular events with frequent negative feedback? Some kind of buffer against external shocks or genetic changes? And what behaviour could one expect of organisms where the cellular interactions are so far-reaching if the cells were oscillators? Waddington suggested that the algebra needed to develop Theoretical Biology of intracellular oscillators, and the cells themselves as oscillators in organisms, could not go beyond binary interactions.

Naturally, Waddington examined the relationship between Theoretical Biology and evolutionary theory. With few exceptions he calls for a return to Darwin's original concepts, and not to follow in the vein of neo-Darwinian reinterpretations. Part of the recent history of Evolutionary Biology has not followed this suggestion, largely due to the development of the phylogenetic analysis of molecular evolution and, more recently, of comparative genomics. And, in a way, ignoring Waddington's suggestion has had positive effects on the extent to which we have begun to assert the existence of certain regularities, universal laws in the evolutionary dynamics of genomes, as well as expanding our knowledge of the tree of life.

Koonin's statement, referred to earlier, continues to confirm these two considerations. But Waddington argued that although we understand evolution better, we still do not have a theory of life. We do not take into account the physiology of the cell and organism in evolution. He proposed going back to studying phenotype variation in all its dimensions, as Darwin believed, and that although genotype variation is a fundamental result of randomness (described most completely by Koonin), laws directing such change do not necessarily apply to variation in phenotype. *Survival of the fittest* means understanding the reasons why some organisms are more efficient in everyday life, not just in reproducing more and having more descendants. Were this to be the primary force driving evolution, says Waddington, it is likely that the biodiversity evolving on the planet would be reduced to organisms consisting of mere bags of eggs and sperm, like some parasitic worms. And this is not what we observe.

According to Waddington, when studying evolution, we should change the emphasis and not focus so much on the fact that some species come from others, like the intimate nature of the adaptation of organisms that, as stated above, is simply studying phenotypic variation related to the ability to live, which in modern terms would be to unravel the physiology of the cell or organism. Similarly, he says, a theory based on the variation in genotype effectiveness, which relies exclusively on maximising offspring, falls short when it comes to explaining how and why living organisms have come to be classified separately in taxonomic terms. It is only now that we are beginning to realise the enormous molecular, genetic and functional complexity associated with the appearance of most of these discrete units in the tree of life we call *species*.

Waddington is critical of the scope of molecular phylogenetics, which was just beginning to develop at the time he wrote about it. He states that:

> So far as I can tell up to date, studies of this kind have no more (and possibly even considerably less) to tell us about the general theory of evolution than do studies of any other phenotypic characters, though they may have something to tell us about the way in which particular proteins carry out their biochemical operations.

As stated earlier, current comparative genomics has been able to make generalisations about how the evolution of genes and genomes occurs, as it has highlighted the importance of proper protein folding—their functionality—to understand the magnitude of the variation in the genes that encode them. But Waddington wonders whether this somehow helps us to understand phenotypic variation, adaptations or speciation. We have reached certain generalisations, but we still lack a theory that accounts for how the above phenomena occur. Andreas Wagner (2011) suggests that, basically, the most important thing in evolution is to understand the nature of innovation, and somewhat surprisingly, formulated a theory of transformative change in organisms, which draws attention to the same points that Waddington said were necessary for a general theory of evolution. In other words, the premise that Wagner formulated to account for the origin of evolutionary innovations, his theory of innovation, would be the explanatory core of the general theory of evolution developed by Waddington. But what would such a theory include? Would it be a theory about the origin and evolution of phenotypes? I think it would. Wagner's theory of innovation should: (1) consider the explanation of how biological systems can preserve existing phenotypes while exploring myriad new ones. The term *biological systems* is the modern way to refer to biological hierarchical organisation, ranging from the molecular components of the cell, the cell itself, or organisation of the organism; (2) unite the different levels of biological organisation; (3) be able to integrate the combinatorial nature of innovations; (4) integrate the fact that the same problem can be solved by different innovations; and (5) consider how the environmental factors influence the dynamics of a particular innovation. As can be seen from these points, Wagner argues that we are talking about a theory of phenotypes, and explains how the different systems that they comprise—biological hierarchy systems—can explain their existence, and the existence of many other phenotypes that will emerge. In other words, a theory of general evolution is a theory about the emergence and evolution of phenotypes.

Wagner's reference to the role played by the environment in his theory of innovation is outlined by Waddington, at the time, as an important gap requiring investigation in the endeavour to formulate a general theory of evolution. Curiously, two major areas of recent research have focused on this issue. Waddington points out, first, the relative importance that the dynamics of small populations can have on evolution. In this regard, surprisingly, the research by Michael Lynch (2007) shows the relative importance of population size on the ability of organisms to generate evolutionary novelty: the smaller the population size, the greater the innovation. Waddington referred to the need to investigate the role of small populations or demos. Lynch notes that the multicellular eukaryotic world has evolved in demos of this nature, in sharp contrast to the evolution of the unicellular world, both prokaryotic and eukaryotic. And secondly, also interestingly, Waddington refers to the need to research biological communities. Species evolve in interaction with each other; not just simple interactions, but really as complex as those observed in any given ecosystem, on identifying the relative abundance of species and their distribution in the trophic pyramid. Is Waddington proposing a theory that integrates ecology with evolution? I believe so, and in fact now we are using procedures provided by the theory of complex systems to measure the interaction of species in multi-species communities, and assess how protected they are in communities with more or less biodiversity, as well as the nature and frequency of interactions therein: symbiotic, parasitic, mutualistic, and so on (Bastolla et al. 2009).

As I have already remarked, Waddington coined the terms *epigenetics* and canalisation to describe interactions between genes, and between genes and the environment, which occur throughout the development, and the buffering that the phenotype undergoes throughout its development. Currently epigenetics refers to the study of heritable regulation of gene expression, without changes in the nucleotide sequence. But Waddington is also the father of the concept of *genetic assimilation*, the process whereby a phenotypic response to an environmental factor is assimilated—through natural selection—by the genotype, so that it becomes independent of the environmental inducer. All these concepts, usually associated with the explanation of certain experimental results, have become key concepts for Theoretical Biology in our time. All three serve as a link between genotype and phenotype, and help us to understand the complexity of genotype-environment interactions. Indeed, these three concepts have been the basis for intensive research conducted over the last decades aiming to discover the in-depth nature of change in phenotypes. They are key components of a theory of phenotypes and, by extension, a theory of the evolution of phenotypes as proposed by Wagner. At the time, Waddington considered such concepts as achievements of Theoretical Biology, and true to say that modern science is putting them to good use.

Dynamics throughout phenotype development would appear to indicate the existence of their canalisation, independent of genetic and environmental contingencies, which leads us to a final point: an attractor consisting of the phenotype of an organism. There are manifold epigenetic trajectories, and they all retain discontinuity in their final products, while they are the product of natural selection. But according to Waddington there is also the archetypal trajectory: a dimension of phenotype evolution on a much larger scale than the time required for organism development. This is the phylogenetic dimension. For Waddington the archetypal trajectory involves what one might call the large body plans of an organism. Without having an entirely clear understanding, he ventures that the archetype trajectories, like the epigenetic ones, are channelled, buffered against changes, and respond to thresholds. But if these thresholds are the product of natural selection, it does not seem so for the archetypal trajectories where certain thresholds are set by the very organisms' internal constraints and which, somehow, delimit the easiest way for the great archetypes to evolve once they appear. The whole programme of current Evolutionary Developmental Biology has been shaped, at least conceptually, around these ideas. True to say, genetic scrutiny of organism development has contributed decisively to consolidating the phenomena of the epigenetic and archetypal or phylogenetic trajectories and their respective canalisation. But Waddington also wondered whether we would come up with a theory able to predict the phenotypic effects that could arise from small changes in any of the thresholds for both types of trajectories, or a theory to explain why some phyla are possible but not others.

Thus far I have referred to a whole conceptual package that was to foreshadow Theoretical Biology, in which Waddington happily became its forerunner and advocate. Furthermore, this bundle of concepts also has interesting philosophical implications, which he examines and compares with other scientific contributions to philosophy. First, we should note that the critical reference Waddington addresses to philosophers regarding the alleged irrelevance of philosophical phenomenology in the context of the living, has been fully justified by the subsequent appearance of a philosophy of substance biology. In fact, it has not just been about the emergence of a particular Philosophy of Science, focussing on the study of Evolutionary Biology, but also because the phenomenology of life has gained importance in the study of the general philosophy of science. Waddington uses two examples to illustrate the philosophical significance of Biology. The first is that Biology provides a dimension of evolution that can explain the fine-tuning between what exists and what can be known by man as the subject of knowledge. Seen as products adapted to the world, it can be argued that species, particularly ours, show a degree of compromise or adjustment between what is perceived and the reality of what is perceived; in other words: the entities perceived are real. In a single expression, natural selection is the only force driving adaptations; it is what Biology brings to epistemology. The second example is an explanation of ethics. The sociobiology theory had not yet appeared. For Waddington it was clear that our species has achieved special heights in its evolution where information is acquired, transmitted and evolves much more efficiently than that associated to DNA. Obviously this is cultural evolution. But such a development, which involves a specific receptor's uptake of a message issued by an issuer, may not be processed or assimilated without the aid of an alleged social authority that enables the receptor to accept it.

I believe that all Waddington's concepts, discussed thus far, form the basis of a high-flying Theoretical Biology.

# References

Bastolla U, Fortuna MA, Pascual-García A et al (2009) The architecture of mutualistic networks minimizes competition and increases biodiversity. Nature 458:1018–1020

Koonin E (2011) The logic of chance. The nature and origin of biological evolution. Pearson Education, New Jersey

Lynch M (2007) The origins of genome architecture. Sinauer, Sunderland

Waddington CH et al (1976) Hacia una biología teórica. Alianza Editorial, Madrid

Wagner A (2011) The origins of evolutionary innovations. Oxford University Press, Oxford

# Chapter 5
# General Systems Theory and Systems Biology

**Abstract** We should not imagine that Systems Biology was born yesterday, indeed von Bertalanffy was a well-known supporter of the need to approach biological entities from the notion of a system. There is a tradition, dating back to early times, calling for the study of living things as a whole. Holism is no stranger to modern Systems Biology and, likewise, von Bertalanffy supported the quantitative study of the whole from the interaction of its parts.

Nowadays, when referring to the term *Systems Biology* we think of a discipline in its own right, which roughly dates back to the year 2000. Systems Biology aims to combine and integrate knowledge from various subjects. Firstly, genomic science in all its forms, ranging from genome sequences to all possible ways of understanding the complete functioning of the cell. Secondly, we have high-throughput experimentation on cells, tissues or organs where we endeavour to control hundreds of variables under certain conditions over space and time. And finally, we also have computer science applied to studying and simulating living beings. There is little doubt that around the same year, the status of investigation was well-developed enough to allow for the realistic modelling of the cell, or parts thereof, or a wide range of complex biological phenomena above the cell in the biological hierarchy. But historical amnesia in science tends to be proverbial if we believe that great ideas spring up overnight without relevant precedents. On the contrary, the fact is that there is frequently prior conceptual and empirical tradition supporting such ideas. Another issue is the particular moment when a situation matures enough for something radically new to be achieved, or a fundamental explanation to appear. Indeed, I argue that this should not detract from such precedents that have been able to push a new discipline ahead or, better still, make some new critical explanatory theory. But that concerns the dynamics of science.

© The Author(s) 2015
A. Moya, *The Calculus of Life*, SpringerBriefs in Biology,
DOI 10.1007/978-3-319-16970-5_5

## 5.1  Systems Biology in the History of Biology

Systems Biology did not spring up out of thin air. On the contrary, it grew from some of the most high-profile developments in Biology of all time. It is said that current systems biology aims to be holistic. How is it possible to ignore this concept, which has been lost in the mists of time, with its particular conception and approach to the study of living things? This issue appears recurrently throughout this book and as I will have an opportunity to comment in the final chapter, in a way current Systems Biology tries to bury the hatchet between holistic and reductionist approaches to the study of living beings, and to realise what I call Goethe's dream. Therefore, to a certain extent we can argue that Systems Biology is a biological science that, for the first time, seeks to study the organism from a global, unitary perspective. This purpose has been a tradition underlying Biology of all time. Probably current Systems Biology goes beyond such holism in that it is able to enter the cell in greater depth, providing a key to the black box that was locked to the previous advocates of biological holism.

A relatively recent precedent of biological holism was von Bertalanffy (1976), whose original work was published in English in 1968, and who is often cited as a theoretical predecessor of Systems Biology. According to von Bertalanffy:

> A system can be defined as a set of elements standing in interrelations. Interrelation means that elements, p, stand in relations, R, so that the behaviour of an element p in R is different from its behaviour in another relation, R'. If the behaviours in R and R' are not different, there is no interaction, and the elements behave independently with respect to the relations R and R'.

As shown, an essential question is the existence of interaction between elements of the system. I think everyone in the world of Biology would be ready to undersign this statement. A different question is to be able to measure the nature of these interactions, and how the elements vary in certain properties that characterise them in as much as they relate or interact differently with other elements. The extent to which a given element does not change the value of one of its properties when it interacts with another variable in the system is what indicates that the component behaves independently, i.e., it does not interact. Indeed, von Bertalanffy also states that:

> The meaning of the somewhat mystical expression, "*The whole is more than the sum of its parts*" is simply that constitutive characteristics are not explainable from the characteristics of the isolated parts. The characteristics of the complex, therefore, appear as *new* or *emergent*. However, if the sum of parts contained in a system and the relationships between them are known, then the behaviour of the system is derivable from the behaviour of its parts.

This quote emphasises that if all parts of a system and all interactions between them are known, then we can predict the behaviour of the whole system. This, however, is not so simple for two reasons. Firstly, because it is unlikely that we can gain full knowledge of all parts of a system; and, secondly, it is even less likely that we can calculate all the potential interactions. But I think that even if we could reach the *Odyssey* of such knowledge, we would still be unable to state the Laplacian *differentiability* of the system. von Bertalanffy claimed that:

> In rigorous development, general system theory would be of an axiomatic nature; that is, from the notion of *system* and a suitable set of axioms propositions expressing systems properties and principles would be deduced.

As I will go on to discuss in greater depth in this book, it would certainly be a great achievement for Biology if biological systems could be formulated as axiomatic theories. But likewise it is true that such systems might exhibit behaviour well beyond that of their axioms or that derived from them. And such unpredictable behaviour would not necessarily be the result of not knowing all the parts and all their interactions in depth, merely a limitation intrinsic to the system itself.

Another issue that concerned von Bertalanffy was *organisation*. In this regard he argues that:

> In Biology, organisms are, by definition, organised things. But although we have an enormous amount of data on biological organisation, from Biochemistry to Cytology to Histology and Anatomy, we do not have a theory of biological organisation, i.e. a conceptual model which permits explanation of the empirical facts.

The author would be shocked if he knew how poorly we consider the so-called "enormous" amount of data available on biological organisation at the time he wrote this. Information on biological organisation, particularly of the cell, has grown by several orders of magnitude compared to that available in the nineteen-sixties, and this is not merely a quantum leap. Holistic concepts have always been steeped in Biology, but have caught on only when they provide a good deal of detailed information which, in turn, enables the formulation and comparison of models shedding light on how life functions at different levels of organisational hierarchy. Therefore, models, systems, organisations, hierarchies, and many other concepts that characterise Biology have always been present in Biology, as I said, but have taken shape when they provide empirical substance. It is the complexity of biological phenomena that has taken time to dissect, and it had to be done to bestow scientific value on all those concepts, which have also been exported to other sciences. Suffice to say, von Bertalanffy was not a mere theoretical biologist; indeed his general theory of systems is general because it recognized that systems are everywhere. However, perhaps his theory was too general at the time he proposed it. He may have perceived that systems were everywhere, but that did not entitle him to claim that he had come up with an explanatory theory, based on the notion of a system per se, which could be a living being, human Psychology, Sociology or Economics. In fact, we are now making progress in the realm of the living, but we still have a long way to go insofar as the mental, economic or social systems are concerned.

von Bertalanffy lived up to the science of his time and as a good theorist, he endeavoured to find the connection between his systems theory and other disciplines and related subjects. In fact, his characteristic eagerness to generalise, perhaps somewhat excessive, is subsumed in his own theory in a number of conceptual developments from very diverse fields which, in my opinion, have been crucial to shaping modern Systems Biology. Let us describe them briefly, not only from the author's integrated perspective as part of his theory, but to the extent that he was forecasting the greatest subsequent advances in future Biology.

For von Bertalanffy, general systems theory was primarily a classical theory of systems that applied mathematics and calculus. But he also understood that, given the complexity of biological phenomena, many of them could not be resolved in complex systems by implementing differential equations, especially nonlinear ones. Therefore he recognized the intrinsic value of computation and simulation for systems theory. We have only to consider the important role computing plays in current Biology to grasp the importance of his intuition.

Indeed von Bertalanffy appreciated the relevance that other more or less related fields had on his general systems theory to the extent that his idea that "systems are everywhere" is reflected in the fact that virtually no biological theory is alien to his systems theory. In his introductory chapter, von Bertalanffy relates his theory to the theory of compartments, set theory, graph theory (with explicit reference to the Relational Biology of Rashevsky and Rosen), network theory, Cybernetics, information theory, theory of automata, game theory, queuing theory and decision theory. von Bertalanffy qualifies all these theories as formulators of various mathematical models. But even verbal models, so common in Biology, also hold a place in systems theory. He states in this respect that:

> Models in ordinary language therefore have their place in systems theory. The system idea retains its value even where it cannot be formulated mathematically, or remains a *guiding idea* rather than being a mathematical construct. For example, we may lack satisfactory concepts of systems in Sociology, but the simple appreciation that social institutions are systems and not a sum of social atoms, or that history consists of systems (ill-defined as they are) called civilizations and obey general principles of systems, calls for a reorientation in the aforementioned fields.

This quote is particularly instructive in two ways. First of all, it again shows the author's undue desire to appreciate that systems are intrinsic to the world, but also recognises that we are often unable to formulate with sufficient precision what such systems comprise. His held a mathematical ideal, but not necessarily the Mathematics that provided the basis of Newtonian mechanics. Furthermore, and no less importantly, this quote shows how von Bertalanffy's concept of system and his notion of interaction are good precedents for what would later become measures of connectivity and interaction in social systems, and subsequently in organisation networks of all kinds, including biological ones.

## 5.2  The Theoretical Method in Biology

In 1977, in an essay reflecting on biological science published in the *Revista de Occidente* Fernando Baquero states:

> So far we have seen how theorising seems to be based on a special kind of biological reality captured by the researcher's mind. This capture -explaining its aesthetic base- is not

quantitative but essentially *qualitative*: the main objects of knowledge are the qualities of reality, or, as we have seen, the qualities likely to come into harmonious or integrated relationships. If the concept of science, since the eighteenth century, has been closely linked to determining quantities and measurements, it may be time to reintroduce -at least on the basis of theorising- the assessment of qualities.

Quality is the relationship. "At the beginning, it is the relationship", says Bachelard. Knowledge is the science of relationships: science probably does not discern things or substances, but relationships. From this perspective, the quantitative concept expresses only relationship functions in qualitative terms: the quantity-quality conflict is thus overcome. Even epistemology talks about the "metaphysics of quality" (of the relationship), in as much as the quality-the relationship-opposes the being: the being is not outside a network of relationships.

Even then, Baquero recognized the need to direct the study of living things from a relational perspective. In fact, for him, *qualitative* and *relationship* are equivalent terms. Why is the relational aspect so important? The author sensed this was probably because the relationship is the key to revealing the answers. So much so, that rather than knowledge of the living being (biological), we are interested in knowledge of the living being in relation to others. Moreover, we probably cannot make a real abstraction of that living being, isolate it, because that would be like missing the chance of achieving real knowledge. A biological being is a living being in that it is immersed in relationships. How much did Baquero foresee on stating that a being is nothing "outside a network of relationships"?

Current Systems Biology endeavours to study the relationships established at different levels of the biological hierarchy, particularly those orchestrated in the subcellular realm. But it has taken us time to the find appropriate procedures to conceptualise the empirical data in relational terms. Once achieved, or underway, what we appreciate is an unprecedented development in the ability to understand what life is. Baquero also advocates that:

> To begin with, Biology has not yet managed to define with certainty all the functions and relationships of its basic biological units (probably because they are not basic), but above all it should be pointed out that both relational forms of evolutionary integration of these units to form hierarchically superior systems, and the keys to the integration of established biological systems remain largely unknown. We will probably need to develop our understanding of this information further before we can speak with propriety of the possibility of a Theoretical Biology, which is currently just one of our great goals of scientific activity.

On the one hand, Baquero recognised the limited knowledge we then had about the roles and relationships of basic biological units and, on the other, how these units interact to produce higher hierarchical systems. According to Baquero, it was still necessary to develop some empirical knowledge to provide content to the relational complex that a living being represents at all levels of the hierarchy. And that is largely what has been happening over the last fifty years, as we have seen a substantial increase in knowledge portentous to the functional units of beings. But at the same time, we have also proposed qualitative formulations that enable us to capture how the interactions between these units, in both the intracellular and the superior levels, systemic and organisational properties arise. The Relational Biology that Baquero called for, Qualitative Biology, is that currently formalized by computer science. At present, the best equivalent to Relational Biology is Biology simulated by computation.

# References

Baquero F (1977) El método teórico en biología. Revista de Occidente (Tercera Época) 19:69–75
von Bertanlanffy L (1976) Teoría general de sistemas. Fondo de Cultura Económica, México

# Part II
# Logic and Computing

# Chapter 6
# Logic, Computing and Biology

**Abstract** Logic and Computing are appropriate formal languages for Biology, and we may well be surprised by the strong analogy between software and DNA, and between hardware and the protein machinery of the cell. This chapter examines to what extent any biological entity can be described by an algorithm and, therefore, whether the Turing machine and the halting problem concepts apply. Last of all, I introduce the concepts of recursion and algorithmic complexity, both from the field of computer science, which can help us understand and conceptualise biological complexity.

The logician and philosopher Quine justifies the need to use non-quantitative techniques in science on the basis that quantitative methods are likely to fail to capture the essence of the phenomenon under study. In fact, not only do the increasing numbers of quantitative variables involved in describing or explaining a given system hinder our understanding of it, but its resolution is impractical, particularly when it demonstrates non-linear behaviour. Quine states (1972):

> Where number is irrelevant, regimented mathematical technique has hitherto tended to be lacking. Thus, it is that the progress of natural science has depended so largely upon the discernment of measurable quantity of one sort or another. Measurement consists in correlating our subject matter with the series of real numbers; and such correlations are desirable because, once they are set up, all the well-worked theory of numerical mathematics lies ready at hand as a tool for our further reasoning. But no science can rest entirely on measurement, and many scientific investigations are quite out of reach of that device. To the scientist longing for non quantitative techniques, then, Mathematical Logic brings hope. It provides explicit techniques for manipulating the most basic ingredients of discourse. Its yield for science may be expected to consist also in a contribution of rigor and clarity - a sharpening of the concepts of science. Such sharpening of concepts should serve both to disclose hitherto hidden consequences of given scientific hypotheses, and to obviate subtle errors which may stand in the way of scientific progress.

I do not intend to defend the use of Mathematical Logic in Biology. Woodger already did so with little success. I believe it is appropriate to emphasise the need for modelling biological phenomena based on qualitative rather than quantitative criteria. And this leads us to consider to what extent Logic and computing languages may be appropriate for modelling and theorising in Biology. Furthermore,

© The Author(s) 2015
A. Moya, *The Calculus of Life*, SpringerBriefs in Biology,
DOI 10.1007/978-3-319-16970-5_6

we should not overlook the close historical relationship between Logic and Computing, as there is little doubt that the former has contributed to the development of the latter. And we could also wonder to what extent some of the illustrious founders of Computing, such as von Neumann and Turing, have contributed to theoretical thinking in Biology and, largely, to ponder on the profound analogies between them.

## 6.1 Turing Machines and the Halting Problem

Let us consider, for example, the idea of the Turing machine. It is an ideal object that manipulates symbols and which, despite its simplicity, can be adapted to simulate the logic of any computational algorithm. This brings us to a new term, i.e., *algorithm*, whose definition is: (a) a finite sequence of instructions that are normally used to make calculations or to process data; and also (b) an effective method by which a set of well-defined instructions carried out or complete a task. That is, to execute an algorithm involves starting from an initial state, going through a series of well-defined states, and ending in a final state that should solve the problem for which the algorithm was posed. A great deal has been written about the profound analogy between how information (software) is processed by a computer (hardware) and how DNA (software) is similarly deployed through protein-assembly machinery (hardware). To read more about this see, for example, Penrose (1991) or, the more instructive, Chaitin (2007).

Nevertheless, I would like to emphasise an issue that should not be forgotten, because it is significant for Biology. Let us assume for a moment that any physical or biological entity is comparable to a Turing machine. This means that this entity can be reproduced (or solved) using an algorithm processed on a Turing machine. Obviously this statement, to the delight of theoretical computing, is highly abstract, but it will be helpful to understand certain properties of physical systems; in short, to understand the world, including the biological world.

The second important question that Turing raised is the possibility that his machine may not conduct or resolve an issue, i.e., it may find a task impossible. And indeed, he found something he called the *halting problem*, i.e., the problem of determining in advance whether a Turing machine, an algorithm, a computer program, would find the desired solution and halt, or continue running forever. If, for example, we want to know if the program will stop after six months, we simply run it and check whether or not it has stopped after that time. The problem arises if we do not define a time limit and try to deduce whether the program will halt without running it. That is, if we can design a program capable of anticipating the results without having to wait and see. I will discuss this matter following a qualitative argument put forward by Chaitin (2007, Appendix I) regarding the halting problem. Suppose we create an algorithm that can decide whether or not a program will halt. In other words, the algorithm would tell us whether the program we have developed will halt or, conversely, enter an infinite loop and never stop. Let us call this algorithm a *halt tester program*. And then let us create a second program that encompasses the halt tester and another program to test it on. This new program, referred to as program two, must have the following feature: if the test program halts, then the second program enters an infinite loop. The interesting question now is: what happens if the test program is the same program two? Let us again recall that program two enters an infinite loop when the test program halts. And now we have the results for program two. If it halts, it goes into an infinite loop, implying that it does not halt; which is a contradiction. And the opposite leads to another contradiction, because if the program does not halt then the halt tester will indicate this and program two will not enter an infinite loop, so it will halt.

More technical, but qualitatively similar considerations were what led Turing to the conclusion that we could not build a general purpose halt tester program. As a corollary of this argument leading to contradiction Turing stated as follows: if there is no way to predict whether a program will halt or not, we cannot do so by reason either. In other words: we cannot rely on an axiomatic system that allows us to derive, from successive deductions from the axioms, whether a program will finally halt or not because, if there is such a system to indicate whether the program will halt or not, leads us to a paradox.

But let us go back now to the preliminary consideration of the analogy between the algorithm and reality, or the ability to algorithmically formulate any property of reality, for example relating to the living world. Let us recall that existing between DNA as an informational entity and algorithm, as if DNA was a huge Turing strip of tape to process on a machine. We can take the analogy even further and imagine that the cell itself is a Turing machine that contains the informational algorithm (DNA) to be processed with more or less complex machinery. Thinking on this analogy, one can conclude that if the DNA is considered as a system of axioms, however intricate it may be, which constitutes cell functioning as a whole, one should assess the significance of a possible paradox in its very core: it is unknown whether the DNA program it is running will halt or not. Could this mean we cannot absolutely predict, for example, the death of a cell? I am concentrating on the issue of DNA and cellular machinery as a whole, but Turing's machine can be formulated on myriad levels, since any structure or function of the cell can be equated to a Turing machine or be Turing-computable, and one may even conclude that the laws of biology are Turing-computable.

To expand on these thoughts I would like to develop a similar line of argument, proposed, prior to Turing, by the logician Kurt Gödel (to consult his reflections and proof on this topic, please see Penrose 1991; Nagel and Newman 1994; Martínez and Piñeiro 2010). He formulated two famous theorems on the incompleteness of arithmetic as an axiomatic system, namely:

(a) *Undecidability theorem*: within a formal system, questions exist that are neither provable nor disprovable on the basis of the axioms that define the system.

(b) *Incompleteness theorem*: "In a sufficiently rich formal system in which decidability of all questions is required, there will be contradictory statements."

When we transfer these famous theorems to the field of natural phenomena, assuming that these phenomena can be assimilated by algorithms, in accordance with Penrose, we will meet with undecidability and incompleteness therein. If the algorithm is the DNA, and we acknowledge, like Jacob in his *Logic of life*, that an organism's blueprint is fully contained within its genetic information thus, we will encounter, in any case, properties of that algorithm that, although known to be true (existing ones for example), are impossible to demonstrate from the laws (or axioms) governing DNA. This theoretical exercise is important, because we could say that such laws are unknown. But what I advocate here is that even though we may get to know them, there will be undecidable phenomena in the system. What is more, if we were able to develop a new law to incorporate to the existing one, we would still have a system in which undecidable phenomena would occur. Besides, there may be the circumstance in which we start off with a system sufficiently rich in its initial axioms, and this has great bearing on how we approach modern Systems Biology. Imagine we can computationally formulate the operation of a cell; something we will discuss in the third part of this essay. The second theorem tells us that inconsistencies will appear, which probably concern mutually contradictory behaviour from the same set of initial statements.

It is also conceivable that life is more complex than Jacob believes and that emergent properties are inherent to their hierarchical organisation, in such a way that the characteristic laws of each level of organisation are independent of its components. If so, the same considerations I made concerning undecidability and incompleteness from the assumption of extreme ontological reduction of life to its DNA would have to be made from the postulates or axioms corresponding to each of the emergent organisational levels of the biological hierarchy.

## 6.2 Algorithmic Complexity and Randomness in Biology

I have already introduced the concept of *algorithm*. Now I would like to introduce two additional concepts: *recursive* and *algorithmic complexity*. They are two fundamental concepts coming from Logic and Computer Science that will help us understand biological complexity better.

The likely increasing complexity of living beings, understood as systematic genesis overcoming certain transitions, may be the product of recursive properties of the objects that make up a certain organisational level of the biological hierarchy. Recursion, or recurrence, is something well known in Mathematics and Computing. When something is defined in recursive terms, we find that the same term that is being defined appears in the expression defines it. That said, it seems we are facing a paradox which must, somehow, be avoided as the defined cannot be contained in the definition. And the way to solve it, both in Mathematics and Computing, is that the defined object contains a narrower version, or restriction, in its definition. Here are some examples. Consider, first, the recursive definition of a *prime number*. We say that any number is *prime* if and only if it is divisible by any prime number smaller than itself, apart from 1. The critical issue that prevents the paradox in the definition is the restriction introduced in the definition, namely: *no smaller prime number*. Because if we say that a prime number is a number that is not divisible by any prime number, we would not understand what a prime number is.

The factorial of a number is another good example, and probably better illustrates my purposes. The recursive definition of factorial is: the factor of any number n is (n − 1)! times n; that is, it is the product of the number in question by the factorial of the natural number that precedes it. So, if we know the factorial of 5 we have (5 − 1)! × 5 = 4! × 5 But now we have to solve 4!. Applying the definition we have 3! × 4, and so on down to 1! to which we assign the value 1, by definition. Therefore, by solving the recurrence for each number less than 5, we obtain 5! = 1 × 2 × 3 × 4 × 5 = 120. As we can see, to solve any factorial function involves applying a recursive definition, which enables us to gradually reach the final solution. We start with the previous number (n − 1), the factorial of which (n − 1)! we have to calculate and multiply by n. But (n − 1)! becomes (n − 2)! × (n − 1). And so on until, as I stated before, we reach 1!.

The factorial of a number also serves as a good example to approach the concept of recursion in the field of Computing. We aim to create a program that computes the factorial of a number n. It is not complicated to use programming language to perform this operation, nor is there only one program able to solve it. But what is important, and common to the languages and programs used, is that the calculation process is recursive, and that the register containing the factorial value changes gradually or is replaced in successive steps by different values until the calculation is complete and the program ends with the register containing the desired number. Indeed, a function is called recursive when the corresponding algorithm in which it is incorporated, using a Turing machine, enables us to calculate the values of the function for its corresponding arguments. The algorithm halts when all values of the function have been executed, and we obtain the final value.

To calculate a factorial does not seem complicated and we can establish a suitable recursive function for the same. But, as already shown above, we can find algorithms that do not halt. Depending on the recursive function that we have posed, it may remain indefinite for some of the values. These functions are known as *partially recursive*.

To discuss the concept of *algorithmic complexity* I will refer to the excellent and classic text by Küppers (1990) *Information and the Origin of Life*, as well as his approach to the problem of randomness in Biology. Randomness can be studied particularly well in nucleic acid sequences, which can be translated into a sequence of binary digits without difficulty. Let us also consider the assumption that DNA, like large software, may be the program able to define the cell.

The idea underlying the point I am endeavouring to show is: if a sequence is completely random, it will probably not be possible to define any law or regularity that can explain it. In other words, we will not find any program that can reproduce it. Conversely, if the sequence in question is not random, we can devise a program that generates it, a program that will have less associated information than the sequence it generates. For example, let us consider a consecutive four-digit sequence of ones and zeros (1100) up to a total of 100 sequences. We could devise a program to indicate to the computer to write the sequence as follows: "Print '1100' consecutively n times, where n = 100". The program in question has a substantially lower amount of data than the whole set of sequences. In other words, we have devised a program that, in compressed form, is able to reproduce the sequence of digits. This could not be done in the case of a completely random sequence, as there is no program capable of generating it from less information. Therefore, the degree of incompressibility of any given sequence is a characteristic of its degree of randomness. A sequence of figures is random when the algorithm required to generate this sequence is similar in length to the sequence itself.

Let us consider a computer C, on which we implement a partially recursive function that is part of a given program p. The program output is a sequence of bits S. The algorithmic complexity of this sequence, $K(S)$, is defined as the length of the shortest program, or $min$ $(C(p) = S)$ able to reproduce sequence S. Once again a sequence S of n digits is random when its algorithmic complexity $K(S)$ contains approximately the same number of bits as sequence S. If we ponder on this a little, we will realise that it is relatively easy to show that a sequence is not random; we simply have to devise a program that generates a shorter binary sequence than the initial sequence. But the opposite is another matter; namely to show that a given sequence is random is the same as showing that there is no algorithm capable of compressing it. This issue is dealt with by Chaitin in his *randomness theorem* (theorem for random reals, see Chaitin 2007 for example). The theorem shows that the randomness of an algorithm, and therefore the property of being a minimum algorithm, is not demonstrable. Within the theoretical context of algorithmic information we have mathematical proof for the following statement: we cannot know if we are in possession of a minimal formula (theory) by which we can predict the phenomenology of the real world. If we ponder on the quoted theorem, we realise that it is directly related to Turing's halting problem and, particularly, with Gödel's incompleteness.

If we revisit the issue of the randomness of DNA sequences, we find a really interesting question if we apply Chaitin's theorem: there is no way to prove or demonstrate their randomness. Obviously, it is an issue of epistemological nature, the bounds of knowledge, because we are not saying that they cannot have that characteristic, but simply that we cannot prove it.

# References

Chaitin G (2007) Meta maths. The quest for Omega. Atlantic Books, London
Küppers BO (1990) Information and the origin of life. MIT Press, Cambridge, Massachusetts
Nagel E, Newman JR (1994) El teorema de Gödel. Editorial Tecnos, Madrid
Penrose R (1991) La nueva mente del emperador. Biblioteca Mondadori, Madrid
Piñeiro G (2010) Gödel ∀ (para todos). El teorema matemático que ha fascinado más allá de las ciencias exactas. Ediciones Destino, Barcelona
Quine WV (1972) Mathematical logic. Harvard University Press, Cambridge, Massachusetts

# Chapter 7
# Life

**Abstract** This chapter examines Conway's cellular automaton *Life* in an endeavour to assess whether the simple rules governing its operation shed light on behaviours that help us understand biological complexity. For example, it is clear that those rules can imply phenomena of cooperation and competition that determine—at any given moment—the likelihood that an individual cell will live on in the next generation, or die out. But complex structures also emerge from the objects that comprise the initial game-board, which self-perpetuate within the bounds of the game. Although the game is deterministic, the exploratory universe is as vast as one may wish.

Goodwin (1994) maintains that Biology is closer to being qualitative (this being understood in the sense of thresholds which are reached in order to determine activity, movement, etc.) than it is to being quantitative. In this regard, he coincides with Quine, whom I have mentioned in relation to the same matter in the previous chapter. We cannot rule out that a considerable number of quantitative models lead us closer to description rather than explaining the structure, function, and evolution of biological complexity. In any case, quantification is not compatible with the study and the genesis of complexity. It has its remit and possibilities, whose influence probably increases as the modelled systems become more delimited. However, this is not the subject I want to examine here, rather, I want to look at the exact opposite, namely: the relevance of studying qualitative models which may allow for a sufficient understanding of biological phenomenology.

The initial hypothesis would be that complexity can arise from a limited collection of simple logical rules containing variables which in turn, allow for a limited number of states. In this chapter and the following one, I am going to firstly examine the question of how complexity can arise from simplicity and how we can demonstrate complex phenomenology through computing. I should also stress that the models I am going to examine (amongst various possibilities) are a particularly suggestive evocation of the behaviour of biological phenomenology. I have chosen two which have appeared over the last 50 years, and the same idea underlies both

© The Author(s) 2015
A. Moya, *The Calculus of Life*, SpringerBriefs in Biology,
DOI 10.1007/978-3-319-16970-5_7

of them: complexity emerges from simple rules, however not all of the rules bring about complexity. The examples used are the computer program *Life* by Conway (1970) and the algorithmic Chemistry of Fontana and Buss (1994a, b).

*Life* was created by Conway, a Cambridge mathematician, and became internationally famous through the Martin Gardner's article in *Scientific American* (October 1970 and February 1971). It involves a cellular automaton with very simple rules, to the extent that it can be run manually to observe how the initial cells evolve over successive generations. Let us imagine a grid, for example squared graph paper. Each box allows for one of two possible states, 0 or 1, occupied or empty, active or inactive, alive or dead. The boxes are to be called *cells*. Each cell is surrounded by eight neighbouring cells, namely: those above and below, to the left and right, and then the four adjacent or diagonal cells. The rules of the game are to do with the active or living state of these eight cells, which determine if the cell being evaluated will be alive or not in the successive generation. The game revolves around only three rules: whether the number of living cells is two, three, or any of the other numbers between zero and eight. More specifically:

1. If two of the neighbouring cells are alive, the initial cell maintains its current status through to the following phase: it remains alive if it was alive and dead if it was dead.
2. If three of the neighbouring cells are alive, the initial cell will be alive in the next generation, irrespective of whatever state it began in.
3. If the number of living neighbouring cells is zero, one, four, five, six, seven or eight, the initial cell will be dead in the successive generation.

The work of Poundstone (1987) which can be used to present Conway's game, shows that the rules entail the idea of interaction between the components of growing populations. The existence of neighbouring cells is important, however, but only under certain circumstances. Indeed, the presence of few living neighbouring cells is insufficient to ensure that the initial cell lives on. It is like solitude leading to extinction. However, extinction is also caused by the presence of many other neighbouring cells because in this case, competition with other cells leads to death. Evidently, the game could have been devised with different rules which might have allowed more life for the initial cell in conditions of solitude and/or extreme competition from neighbouring cells.

However, Conway was probably aware that it was exactly by using an intermediate number of neighbouring cells that added greater richness and complexity to the game as it developed. On the other hand, progress is completely determined by the rules of the game of course, but also by the arrangement and the number if living cells in the initial grid.

Before looking at the variety of structures which living cells can bring about over generations, it would be appropriate to think about what I consider to be an important similarity between the game and biological evolution. It is unlikely that Conway was aware of this matter, because he was more interested in using the simple rules to derive a deterministic world which could ultimately become very complex according to the type of structures it contained and how they evolved.

However, the fact that he arrived at the conclusion that the best set of rules are those which involve an intermediate level of interaction with neighbours (living ones), is something that we should bear in mind. On the other hand, it is vital to specify more about the meaning of the biological similarity of the interaction between neighbouring cells in the game.

Conway establishes that an initial cell will be alive in the next generation if some of its neighbours are alive. However, in order to link the game closer to biological evolution, we can imagine that the cells replicate (or reproduce) or die out in the following generation, as if talking about the idea of generating offspring or not doing so, that is to say, dying. Following rule 1, if two neighbouring cells are living, the initial cell holds its current status through to the following generation: living if it was living (it reproduces) and dead if it was dead. This is a *rule of conservation* or maintenance. On the contrary, rules 2 and 3 are *rules of transformation*. They ensure that a living cell will die, or that a dead cell will live. This statement needs to be explained in order for it to make biological sense. Moreover, rules 2 and 3 are asymmetrical in two senses of the term. Firstly, they differ between the numbers of living neighbouring cells and secondly, taking account of the very nature of the rule which is applied to the state of the initial cell in the following generation. Indeed, according to rule 2, an initial cell which is living can replicate, or if it was initially dead, it can be transformed into a living cell in the next generation. This latter condition is particularly interesting because it is the circumstance which allows the number of living cells on *Life*'s game board to increase. The other rules involve the reproduction of the initial cell to the next generation if it was living or, simply, its death. To a certain extent, rule 2 allows for the initial cells to multiply. It is better to interpret this rule from the perspective that, by applying it, an increase in initial cells is achieved, rather than holding that the cell has come back to life, when it may have been the case that it was dead. In my opinion, this is effectively the least biologically acceptable condition of Conway's game, because the primary dictum of Biology is that life can emerge from life, but life cannot emerge from death. Some well-known biological phenomena, such as spores or hibernation, could be viewed as similar to these circumstances of multiplication starting from the initial cell's state of death. However, both hibernation and spores cannot in any regard be equated with death. On the contrary, they are both special states of life—suspended life in fact- and only when the organisms involved have suitable environmental conditions, does life make an appearance, and in the fullest sense of the term.

In short, together the three rules allow for replication, an increase or the death of the initial cells from one generation to the next. Rule 3 is a rule concerning the death of cells, given that if the initial cell was dead, it will remain so, and even if it was alive, it will also die. It is very interesting that the condition which determines mortality in the following generation is that the number of living neighbouring cells is either very low (zero or one) or very high (between four and eight). From the perspective of a biological analogy, it can be stated that the rule in question is complex, as the phenomenon determining the death of the initial cell cannot be the same when there are very few or an abundance of living neighbouring cells. As I mentioned previously, in the first case, a kind of death is brought about by solitude and in the second, death is caused by overcrowding. It is clear that the initial cell needs neighbouring cells to survive, and if there are few or none, it will not survive. However, nor will it survive with too many neighbouring cells.

Another interesting question which is very evident in Conway's game is that it is a game which has more to do with death than with life. Indeed, if the initial cell begins as one of two states and its condition of life or death in the following generation depends on the number of living status of the neighbouring cells (from zero to eight living cells), then there are eighteen possible results for the life or death of the cell in the following generation. Three for life and fifteen for death. The probability that the cell will live on into the next generation is not even 20 %. It is probable that these conditions prevent an excessive number of structures from proliferating on the grid, which would be unmanageable for a human observer. It is worth remembering that this game was devised to be played with a pen and paper.

Due to its biological significance, it is necessary to stress the importance of the link between the life status of the neighbouring cells and the status of life or death of the initial cell in the following generation. Turning to the rules, it is notable that the concept which underlines them is *cooperation*. It is not deemed, however, that the aforementioned cooperation can be greater owing to some type of relationship between them. Therefore, the type of relationship existing between the initial cell and the neighbouring cells could be seen as similar to some well-known forms of cooperation in Biology. However, we should examine the consequence of these types of cooperation on the potential survival of the initial cell. I have already been able to outline that if cooperation with neighbouring cells is low, which is measured by the fact that there are few living cells to coexist with, the initial cell will die. It is as if there is a minimum threshold of care to be provided by the neighbouring cells. If this threshold is not reached, the initial cell will die. However, the opposite situation arises when there is an excessive number of neighbouring cells. A biological comparison to this situation would not necessarily involve an absolutely excessive amount of care provided by the neighbouring cells which kills the initial cell. Rather, it should be thought of as a situation concerning competition for resources, an occurrence which limits the possibilities of surviving into the next generation. Consequently, interaction with neighbouring cells in Conway's game is more subtle and closely linked to Biology than it may initially appear. It not only involves interaction but also competition. The relationship with neighbouring cells establishes a kind of cooperation/competition continuum which, on one hand, requires necessary cooperation so that the initial cell endures. However, on the other hand, competition is not necessarily disastrous when the number of neighbouring cells is limited. This is to say, there is a cooperation/competition optimum which is based on having between two to three living neighbouring cells. Although there may be very little competition when there are few neighbours, it is necessary to have cooperation beyond a threshold level so that the initial cell can replicate. Likewise, there may be sufficient cooperation over this threshold level, yet if there is a very high level of neighbouring cells then we surpass the competition threshold which also determines that the initial cell will die out in the next generation. These considerations are probably very similar to the theories supported by Nowak in his book *Supercooperators* (Nowak and Highfield 2012). For the author, Evolutionary Biology seems to revolve around a permanent dilemma from the outset of life, between cooperation and competition, irrespective of the biological hierarchy involved. Nowak maintains that cooperation is crucial to the emergence of complexity and that a large number of evolutionary developments come about as a consequence of appropriate solutions to the cooperation/competition conflict in favour of the first option. This is what results in the emergence of new levels of complexity in the biological hierarchy. Multicellular organisms are a good example of this as they show how the cells composing them, have at some point in evolution, been able to overcome the conflict arising from competition amongst themselves. Faced with competition which would lead them towards independent and competitive lives, they formed a new and more complex structure involving the sharing of tasks.

Thus, it becomes evident that Conway's game which is ostensibly so simple, seems to be able to capture fundamental properties of living organisms. I have referred to concepts such as generation, replication, multiplication, death, cooperation and competition, which are present in the game's dynamics. As such, it is necessary to acknowledge that the behaviours exhibited over the course of the game, may have some similarity with those shown by living organisms over their evolution. Equally, there are other important aspects which do not feature in the game. The cell only allows for one of two possible states, and we know nothing more about it. It is a real mystery, somewhat like Leibniz's monads, it cannot be boiled down to different parts. It is an indivisible unit that only allows for two states, which are totally determined by the states of the neighbouring cells and not by its own intrinsic properties. The cell possesses no capacity to genetically modify itself so that it might have a greater chance of survival than other initial cells in the grid. The cell is living or dead. If it is living, it will live on into the next generation two out of every nine times, depending upon its neighbours. If it is dead it will only live on once every nine times.

## 7.1 The Development of Complexity and Emergence

I would highly recommend that interested readers look at the work of Poundstone (1987) in order to examine the types of structures in Conway's game, how they evolve, generate new cells, and move or interact. In this regard, a starting grid full of living cells will die out in the next generation, as would a starting grid with a single living cell. In order to make the game interesting, so there may be something surviving into future generations, keeping its form or transforming into something new, it is necessary to begin with structures having a minimum number of living cells. Nor will two living cells succeed: they will die. We may begin with three cells. For example, three living cells arranged horizontally will lead to three more in the next generation arranged vertically. In turn, the following generation will have another three horizontally. As two generations are needed to repeat the structure, the frequency is two. There are structures with a larger frequency which is several generations. There are others whose frequency is zero. For example, four living cells arranged in a square repeat their structure permanently in each successive generation. Others which do not have a frequency are unstable, in so much as they begin a process of change and end up forming a new structure as generations pass, a structure which in turn can be periodic, but which is different to the original structure.

There are structures with five living cells which are very interesting. Conway named one of them *glider*. This structure has the interesting ability to move in the grid, and after four generations, it repeats the original structure (it is also periodic), however it will have moved in the grid in relation to the original position. I refer to it precisely because of this ability to emulate an important biological phenomenon: movement.

Another interesting structure with five living cells is *pentomino*. It is shaped like a small letter *r*. What is exceptional about this structure is its capacity to generate new structures over time. If it begins with a sufficiently large grid (this matter is of particular relevance as will be shown later), the pentomino is capable of creating a lot of life. Writing about this structure, Poundstone shows that if the game is started with a single *pentomino*, at the end of 1103 generations a grid is reached which contains 25 of the aforementioned structures and others, which have 116 living cells. These structures stabilise at this generation, and whilst some may move or change, considering that they are periodic, the fact is that they no longer have any kind of interactions with the rest following this point. They all remain as if frozen in the game's time. On the other hand, it is surprising to note that this generation does not amount to the point of greatest diversity. In generation 821 there are 329 living cells distributed in a series of structures. This is certainly an interesting explosion of diversity; however it also leads us to think about the possibility of a moment when the initial structure is no longer able to generate. This also has a biological comparison. It is recognised that some evolutionary radiations are very successful and capable of systematically creating new species over time; however, others are barely capable of this. There is also the case of phylogenetic depletion, in the sense that evolutionary radiations which are successful for a certain amount of time, subsequently stop being successful at other times. Given that there are structures which have the ability to be creative over time whilst others disappear or simply repeat, we must arrive at the conclusion that they possess different intrinsic properties. Compared to other structures, pentomino possesses a greater capacity to evolve creatively. These are inherent properties and are completely determined by the initial configuration and the game rules.

However, there is another factor which we must mention, and which also has a biological analogy. This factor is the grid itself, given that it functions as a space where pentomino can display the said capacity. If that space is limited, it is likely that neither maximum biodiversity nor the subsequent stabilisation will be detected. The example that Poundstone provides involving pentomino requires a grid which contains around 6500 boxes (more precisely $57 \times 114 = 6498$). Had a smaller grid been used at the start, some structures and the interactions amongst them would not have appeared. Moreover, it is necessary to consider the pentomino's starting position on the grid. Conway established that it was necessary to have a grid such as that previously mentioned in order for the structure to develop all of its creative potential. However, this was subject to the initial placement of the pentomino. He did not place it in the very centre of the grid, but rather somewhat lower down, in the lower left-hand side area. If, for example, it was placed at either of the ends of the grid, we would probably be left with an effect similar to that of having a reduced grid, namely: the pentomino would not develop all of its potential and would be truncated at some point during game time. Therefore, it would appear that available space and the initial position are important factors which contribute to a greater or lesser extent in facilitating the structure's evolutionary development. The biological comparison to the grid and the structure's initial position is evident. The grid not only constitutes a physical space and the ability of organisms to find new niches, it also represents the resources available for survival, or even facilitates encounters and interactions between organisms. The starting point bears an interesting similarity with the so-called adaptive peaks and valleys. There are well-documented cases in organic evolution concerning the impossibility of organisms reaching optimum adaptation points by moving from optimal places to other universal ones. The incapacity to reach them is due to the fact that organisms must pass through valleys in the adaptive landscape. If they are placed in an optimal adaptive position, then crossing a valley means a loss of efficiency, something which cannot be the case when we consider that natural selection cannot contribute to this loss. Only random circumstances or landscape alterations could allow such optimums to be reached.

Pentomino itself can be used to demonstrate another interesting property of the game and its similarity to Biology. This concerns the encounters between structures and what happens in future generations following these encounters. It should be noticed that rules 1–3 of the game refer to the relationship between the initial cell and its neighbours; I have classified such forms of interaction as being either cooperation or competition. However, we are now considering interaction at a higher hierarchical level, given that what we have are structures, comprised of living cells, which, through their generational dynamic, are able to encounter other structures. However, there are no new rules relating to these. In other words: objects emerging from the world of *Life* exhibit interesting behaviours which are shown in their fullest reality when the aforementioned structures evolve alone in a sufficiently large grid. Some structures are creative, others repeat themselves with a certain frequency, and others remain the same. In this world of solitary structures, we notice that they remain as such (zero frequency) or that they repeat: they are stable structures over time. Can we consider them to be organisms made of cells which replicate themselves one generation after another? We can to a certain extent. However, this world of solitary structures bears little relation to the reality of the biological world where interaction amongst organisms is the norm rather than the exception.

It is precisely in the interactions between structures where can best be seen the difference which exists between the basic rules governing the component cells and all the structures they generate, and what may be called rules associated to new structures or *emerging rules*. A property is said to be emerging when it is not predictable based on the simple application of rules which work for component parts of the object or structure in question.

It is worth considering the result of encounters between two or more structures in any generation. As the encounter does not have emerging rules, the fate of the relationship depends on the underlying rules 1–3. As if absorbed by the context defining such rules on the grid, the structures which meet form a new structure, which is itself subject to the same rules as the individual cells which compose it. In this way, a certain structure is then produced in the following generation. It seems that Conway's structures lack autonomy or individuality, when these properties are understood as the ability to remain the same irrespective of the type of structure encountered. Are any of Conway's structures completely immune to the actions of another? I do not know of any. However, how can we rule out the possibility of finding one which possesses the property of autonomy? It would be important to be able to be demonstrate this as a mathematical possibility. I am also unaware whether this has been achieved. On the other hand, it has been possible to demonstrate the existence of structures which cannot be the product of any other. These are the so-called *Garden of Eden* structures as they have no predecessors.

## 7.2 Determinism and Closed Evolution

Determinism in *Life* is absolute, with three fundamental conditions which inexorably shape its subsequent development: the size of the grid, the initial structures used to start the game, and rules 1–3. Of course, there is no possibility of hoping for changes to emerge from this. If the process if repeated with exactly the same conditions we will obtain the same answer. Even so, the game offers an immense range of possibilities to explore. Consider for example, the grid size. If the initial space is different, there is the possibility of having different evolutionary stories. Of course, this also happens if we use different initial structures on the same sized grid.

In general, with initial structures that are adequately creative and in a sufficiently large space, we tend to observe evolution which leads to a balanced generation. This generation will contain a given number of final structures with a low number of cells. They will be isolated with a frequency of zero, or with some fluctuation. The structures in question freeze over time, in the sense that they remain the same and make copies of themselves after each generation, whilst remaining apart from one another. The fact that the world of *Life* is restricted by the aforementioned conditions of space, initial placement and classification of structures, as well as rules, provides the chance to observe a kind of closed evolution. If no uncertainties are added to the system then we have an end system. On the other hand, I have already mentioned the fact that *Life* does not constitute a world which allows for the emergence of structures with autonomy or individuality, which are capable of maintaining their status when they interact with any other type of structure. Unfortunately, nor does the game take into consideration any type of mutational rule, which could allow for the creation of structures over time, beyond those which can gradually emerge as a consequence of the initial rules being applied. Is it worth considering the feasibility of having an initial structure or series which may allow for unlimited growth? In the previous section we already saw how a structure as simple as pentomino, alone, can evolve for 1103 generations. At that moment, the constellation of created structures remains isolated, and lacking the possibility of new interactions, the structures freeze over time. Although it is only conjecture, the answer to the previous question is probably no. There are conditions which would support this conclusion. The first is space. It is finite. Many of the simple structures which emerge in the process move towards the edges of the grid and remain there isolated and frozen either because they have the ability to move or because they are the by-product of interactions which displace the cells in the grid. Secondly, it is possible to inductively reason that all cases coming from structures or series of them which have a certain creative dynamism, will die sooner or later.

However, there is a characteristic of *Life* which is reminiscent of biological evolution. Three main types of structures are classified in the game. Firstly, there are the *stable* structures which emerge as the consequence of the past action of others which end in them. These structures which are isolated then generate copies of themselves. This type of stable structures in isolation is the product of the convergent evolution of other structures. A second typology has to do with oscillating structures with a determined frequency. Once one of these appears, it gives rise to another in the following generation, and after a certain number of rounds, the initial structure is repeated. To a certain extent, these structures are as stable as the first category, however it is necessary to call them *periodically stable*. It is worth stating that these structures are a product emerging from others, irrespective of the period when they encounter one another. In so much as the period is temporary, one can think that a good analogy with Biology would be the life cycle of organisms. An insect with a complete life cycle goes through the phases of egg, larva, pupa and adult, shall we say four periods. The problem with the analogy resides with the convergence of each structure. In *Life*, many structures can give rise to each of the periodic structures, whilst in biological cycles, this possibility is restricted given that what primarily determines the development of the organism is the ontogenetic development of the genotype. Indeed, the cycle is strongly restricted by individuality itself, something which seemingly does not exist in *Life*.

As can be deduced from the two main types of previously named structures, there is a third type made up of the real unstable structures. They have no ability to support themselves and are those which evolve towards stable or periodically stable structures. Finally, there are the unique or solitary structures. I already had the opportunity to mention them with the example of the *Garden of Eden*. These are not the product of any previous structure. That is to say, they do not originate from a previous unstable structure. They themselves are unstable structures. A biological equivalent for these structures could be made with the so called *frozen accidents*; fortuitous events in Evolutionary Biology, which are unique, and to a large extent have been involved in important evolutionary innovations or inventions. These have subsequently given rise to a plethora of biodiversity at different levels of the biological hierarchy.

*Life* would appear to lack a shielded structure, an object which is able to remain the same over time irrespective of whatever it interacts with. Its stability would be in my opinion, greater than the previously mentioned stable structures. They are stable in so much as they are isolated. My theory is that whilst there is no emergence option in *Life,* this type of structure is not possible. For it to be feasible, it would be necessary to introduce new rules into the game. Let us consider to what extent the second of the approaches, the algorithmic Chemistry of Fontana and Buss (1994a, b) based on lambda calculus, allows not only for the emergence of individuality, but also of an open evolutionary universe, something which is so vitally important for Biology.

Readers will have noticed the enormous exploratory possibilities offered by cellular automatons. Conway's game is very basic. Only by varying the starting rules, for whatever motives, will we encounter new automatons which will provide new patterns. There is always an underlying idea that simple rules can quite often create complex patterns. There is a book called *A New Kind of Science* written by Wolfram (2002) which is a systematic project, purportedly of great scope, which includes Biology as well as the other sciences. It is interesting to mention here, if only briefly, the meticulous work which the author has carried out in the field of cellular automatons. Wolfram asserts that computational research with automatons is able to provide solutions to well-known problems related to complexity which appear in a diverse number of fields such as Mathematics, Natural Sciences, Social Sciences, Technology, Art and Philosophy.

# References

Conway (1970) The fantastic combinations of John Conway's new solitaire game "life". In: Gardner M (ed) Mathematical games, scientific American, vol 223. pp 120–123

Fontana W, Buss LW (1994a) The arrival of the fittest: toward a theory of biological organization. Bull Math Biol 56:1–64

Fontana W, Buss LW (1994b) What would be conserved "If the type were played twice" Proc. Nat Acad Sci USA 91:757–761

Goodwin B (1994) How the leopard changed its spots: the evolution of complexity. Weidenfeld Nicolson Illustrated, London

Nowak MA, Highfield R (2012) Super cooperadores. Ediciones B, Barcelona

Poundstone W (1987) The recursive universe. Cosmic complexity and the limits of scientific knowledge. Oxford University Press, Oxford

Wolfram S (2002) A new kind of science. Wolfram Media Inc., Champaign

# Chapter 8
# Algorithmic Chemistry

**Abstract** The pioneering work by Fontana and Buss, algorithmic Chemistry, resorts to lambda calculus to simulate entities with recursive properties. This theory is central to computation and is a good counterpoint to the universe of deterministic simulation of Conway's automata. Objects interact with each other and, consequently, other objects—new or not—arise from this reaction and join the group. The possibility of introducing stochasticity and mutability to the objects gives greater biological realism to the system. Not only do stable entities emerge over time but—regardless of the initial composition of objects in the reactor—under certain conditions such entities converge towards stable, self-perpetuating structures. The world devised by Fontana and Buss helps us to reflect on the extent to which contingency and necessity are two equally important terms in biological evolution.

The system put forward by Fontana and Buss (1994a, b) is interesting because of the way in which it emulates important properties of biological systems. Can we consider it to be a kind of evolutionary grammar which, when made to work, simulates the phenomena of biological evolution such as replication, variation, multilevel organisation or complexity? As with *Life*, the system devised by Fontana and Buss is of interest not only for its philosophical implications, but also as a basic theoretical approximation to biological complexity. In a certain sense, it is a qualitative biological theory. Simulation is used to obtain results which are approximations to the reality of the biological world, especially if the studies are carried out computationally. The concept of *algorithmic Chemistry* arises from a combination of two terms. The term *Chemistry* is due to the fact that objects in the Fontana and Buss system interact as if they were carrying out a chemical reaction. That is to say, two objects (or molecules) react and produce a new object. The reactions do not adhere to the real principles of chemical reactions, such as the conservation of energy and mass. These are abstract chemical reactions. The *algorithm* concept refers to the fact that the authors created a program, which is described concisely and informally below:

1. In the random creation of an initial series of objects from the reactor, objects are transformed into reduced versions of their normal shapes. These shapes are

© The Author(s) 2015
A. Moya, *The Calculus of Life*, SpringerBriefs in Biology,
DOI 10.1007/978-3-319-16970-5_8

obtained by applying a series of rules which lead the initial objects to a point
where they can be reduced no more.
2. The objects collide one against another, following rules. Each collision creates
   new objects which are processed through axioms and conveniently reduced.
3. If a new object is obtained following the collision, it is incorporated into the
   series, and one of the prior objects is eliminated at random. This means that the
   number of objects in the system remains constant.
4. The reactor carries out as many reaction cycles (or generations) as are wanted,
   and a series of objects may gradually appear which exhibit new and complex
   patterns (for example self-maintenance) in relation to the initial objects.

What is interesting about the algorithm is that the authors find complexities in
the reactor which are *reminiscent* of the fundamental properties of Evolutionary
Biology, such as replication or the stable organisation of complex structures, or
even levels of integration of these structures into even more complex structures,
thus creating something similar to the kind of organisational hierarchy held in
such high esteem by biologists. Fontana and Buss have developed a theory of biological organisation based around the behaviour of abstract objects which interact
following the rules of λ-calculus (lambda), a fundamental type of calculus in the
field of Computer Science.

One of the properties which it emulates is self-maintenance. In Biology, self-
maintenance is understood to be the property of a biological entity (for example, a
cell or organism) which manages to maintain itself by letting a series of chemical
compounds enter inside, with subsequent metabolic transformation. Properties associated with the self-maintenance of these entities are, amongst others, resistance to
disruption, wave extendibility and historical dependence. Fontana and Buss created
a suitable theoretic framework primarily based around the premise that the components of a self-maintained object interact with others in such a way that this interaction results in the construction of a new object. According to this, self-maintenance
is a form of feedback which emerges as the consequence of the type of properties
possessed by the components of a certain object: properties which allow it to maintain itself. The specific working relationships between the components, which col-
lectively ensure the regeneration of an object, can be defined as *organisation*. In
this regard, a theory about organisation is a theory about self-maintained objects.
Fontana and Buss develop this theory of biological organisation using a type of calculus which is fundamental to computing, and whose basic principles are based on
the nature of certain objects and their recursive properties.

## 8.1 The Reactor and Experiments

There are two basic abstract ideas which are fundamental to any theory about
biological organisation. Firstly, Biology can be considered as *constructive* at least
at various levels of its hierarchical organisation. The term constructive is intended

to indicate the capacity of biological entities to generate new entities when they interact. Secondly, interactions can be grouped into *equivalence classes* in that different interactions are capable of creating identical and stable entities. As Fontana and Buss (1994a) already understood, one of the best comparisons in computing to formally describe the behaviour of biological entities is lambda ($\lambda$) calculus. This calculus formally collects the two prior properties: constructivism and equivalence classes of biological entitites. Lambda calculus was developed by Church (1942), one of the people who introduced the notion of the *computable function*.

Lambda calculus plays a very important role in modern computing theory. Indeed, structures made using it can offer great intuitive insights into the semantics of programming languages and their development. Lambda calculus can be considered as a paradigmatic computing language, because, despite its simple syntax, it is powerful enough to describe any function which is mechanically computable. Lambda calculus considers the functions being processed as rules, determined by programs for *machines applied* to or *operating in* any other program. The objects captured by this calculus are simultaneously *argument* and *function*. More specifically, any function can be applied to itself.

The computing experiments carried out in the reactor, are based on the axioms of calculus. There is an explicit similarity to be made between the types of interactions which can eventually take place between simple biological objects and interactions produced between $\lambda$-expressions. Lambda expressions are treated as if they were physical objects. The world of these objects is interpreted according to the rules of $\lambda$ calculus, which defines the nature of the objects and how they can be syntactically transformed. The initial world is a series of different objects in their reduced normal forms. In such a world, the basic event involves the interaction or physical collision between two objects taken at random. The collisions take place between two objects chosen at random. After the collision, the newly created object is reduced to its normal form, and is subjected to various syntactic and/or functional restrictions, named filters, which to a large extent are related to the types of experiments we want the reactor to carry out (see below). If the object passes through all these filters, it is added to the system, after which random elimination of another object takes place, so that the number of objects throughout the experiment remains constant. Fontana and Buss carried out three main types of experiments, which they called level 0, 1 and 2, respectively.

They typically started the level 0 experiments with 1000 different objects in reduced normal forms. Generally, after several thousand generations or consecutive collisions, they observed the imposition of simple objects with the capacity to copy themselves; or the imposition of series of them where the capacity to copy was the consequence of their mutual coupling to form a *hypercycle*, a well-known biological structure. This involves a series of catalytic cyclical reactions, whereby the molecules involved catalyse one another, which means that the hypercycle remains an autocatalytic system.

Moreover, Fontana and Buss also observed that when the system was disturbed by the introduction of random objects at some point, it was those with the self-copying capacity that imposed themselves. The ability to copy, by objects different to the copied object, or to self-copy, can be considered as *reproduction*. This is the primary characteristic of level 0 experiments, which have, as already indicated, a tendency in the long term to create objects that have the capacity to reproduce. The structures of *Life*, also display reproductive abilities, but within a context of much lower interaction than the present one. Remember that when some of the structures in *Life* were isolated, they had a self-maintenance ability. It should also be remembered that others which had a periodicity, bear a certain resemblance to the objects which here display hypercyclic behaviour. However, neither Conway's game nor the level-0 experiments seemed to produce structures or objects with a self-maintenance ability, a trait that is highly relevant in Biology. Fontana and Buss then began to argue that there could be a real possibility of potential antagonism between reproduction and self-maintenance, something which, on the other hand, is fundamental to Biology. Organisms are characterised by possessing a metabolism, independence or isolation in relation to their environment, a reproductive capacity, and evolution. The effort that they invest in, let us say, their metabolism, can compromise their reproductive investment, and vice versa. Therefore, it would not be preposterous to attempt to carry out an experiment without involving the reproductive ability. This is how they outline the level 1 experiments, in which they explicitly included a condition, requirement or filter, which would prevent the ability to copy. Otherwise, the remaining conditions were similar to the level 0 experiments; unlike the results obtained.

The first interesting result is the presence of structures with a self-maintenance capacity. This observation is highly relevant, as it already offers a type of object which did not emerge in either *Life* or in the level 0 experiments. Objects created when the reactor is in an advanced state self-maintain. In so much as these objects, owing to their syntactic configuration, can maintain themselves dynamically, they exhibit a certain sense of organisation, which is evidently reminiscent of biological organisation.

A second average result which is of relevance, are the emerging laws, both syntactical and functional. I want to stress the role of the latter. Fontana and Buss arrive at the conclusion that comprehensive regularities emerge as a result of the collective behaviour of objects that interact locally. Furthermore, they are emerging laws in the sense that they do not make any reference to what occurs at a micromechanical level during basic operation of $\lambda$-calculus. Evidently, we are faced with a universe which produces much more reliable simulations of what occurs in Evolutionary Biology. This is because the emerging systems that emerge gradually throughout Evolutionary Biology are linked to their own laws which do not incorporate, or are independent from, those governing the components of such systems.

Another interesting property observed by the scientists was the robustness of the self-maintained objects. In fact, the elimination of some of their components did not prevent the original structure from restoring itself over the course of the experiment; this trait is very dependent upon self-organisation. The same thing appeared to occur when the system was disturbed with the random introduction of new objects. Unless the disturbance was large, due to the type of object placed in the reactor or because of the high periodicity level of its insertion, the self-maintained objects remained quite robust against disturbances. However, it is worth stressing that this was not incompatible with the capacity to generate something new. In fact, they detected the presence of some disturbances capable of creating a new emerging law, which in turn, coexisted with those that existed prior to the disturbance.

In short, the level 1 experiments showed the emergence of objects with a capacity for self-maintenance (with individuality, which was absent from the structures in *Life*), within a dynamic system of interaction with the rest of the objects. They were subject to emerging laws in their own organisation, which were not a product of the axiomatic system of calculus. They were robust in so much as they were capable of restoring internal components as a consequence of the type of organisation and also faced with disturbances. However, due to certain disturbances, objects with new emerging laws could sometimes appear, these were then added to those that had existed previously. If this summary is read in detail, it is possible to see the striking similarity this type of organisation bears in relation to that observed in biological evolution and organisation.

Finally, Fontana and Buss thought of a third type of experiment. Ultimately, the structures they had obtained with level 1 experiments, logically, led them to think about what might happen if the same self-maintained objects could be used to begin level 2 experiments. Would there be a possibility of creating metaobjects which would be more complex than those detected in the advanced stages of the level 1 experiments? As one might already have guessed, in light of the very dynamics of Evolutionary Biology, they obtained a two-fold answer. In some of the level 2 experiments, objects appeared with the characteristics of those of level 1. That is to say, no significant evolution was detected. However, a large number of these objects did create metaobjects or metaorganisations over time, which were labelled as level 2. These were really complex structures in that they contained self-maintained level 1 objects along with a series of objects which, without being self-maintained themselves, did allow for communication between the level 1 objects, in order to create a maintained object of a higher category than those of level 1. They also observed that the level 2 organisations were very sensitive to initial conditions (the type of initial level 1 objects) as well as surrounding conditions, as well as filters or restrictions.

With regard to the emergence of laws, there is a repeat of the situation observed with the creation of emerging level 1 laws. Laws which are unique to level 2 emerge whilst maintaining those of level 1. Those of level 2 bear a relation to specific actions, which literally stick down the level 1 organisations, as well as new metabolic fluxes, formed by level 1 components, which allow for the union of those organisations, but with principles of organisation which are new or unique to level 2. In fact, variations in those fluxes allow for potential compositional change amongst level 1 self-maintained objects. Finally, Fontana and Buss also studied the effects of various types of disturbances, normally involving the introduction of random objects in various forms over the course of the experiment. Results varied, although it can generally be said that when the disturbances were insignificant, the organisation of level 2 remained unaffected. However, in some cases where disturbance had a more obvious effect on the formation of level 1 organisational levels they were able to detect a degree of simplification amongst the type of level 2 laws.

## 8.2  Contingency and Necessity

Probably because the only Biology known to us is that of our planet and because biological evolution has occurred only once, it seems that it is itself fortuitous and that many of the processes occurring in its history also exhibit this trait once they have emerged. Fontana and Buss (1994a) note that it is important to know what would happen if we pressed rewind on the film of life. Doing so would allow us to gain some perspective on how much *contingency* (or chance, to use Monod's wording) and how much necessity are involved in life. While we are waiting to determine whether life has existed, or does exist, on other planets, what it is like and how it has evolved, the fact is that approximations, such as those discussed in this essay, demonstrate that we should probably strive to understand the necessity factor. Evolutionary Biology has endeavoured to show that contingency, history, have inevitably conditioned its own unpredictable ramification. However, I believe we should ponder about this consideration in greater depth in light of what theoretical experiments indicate, such as those of Fontana and Buss. Personally, I enjoy explaining to students what would have happened in the story of life had the dinosaurs not become extinct. Those beings populated the planet, and many habitats (land, sea, lakes and the air) were filled with well-adapted species. In comparison with the biodiversity of dinosaurs, mammals were no more than poorly evolved shrews which were incapable of ever prevailing over those majestic ubiquitous beings. However, dinosaurs did become extinct. The question of how fortuitous or contingent the process was, is of great interest, and there are various competing models of extinction. However, it is true that our species would not have evolved had they remained alive. Or would they? The question we must ask ourselves here, refers specifically to necessity given that contingency seems very obvious: dinosaurs vanished from the face of the planet in a very short amount of time span. Looking back at this, we have to ask where the evolution of dinosaurs would have led to. Would they have reached the point of developing a language and higher intelligence, like some kinds of subsequent mammals? This is the question of necessity.

In the absence of empirical information, or due to its scarcity, projects such as those by Fontana and Buss help to show, the almost inevitable or necessary manner in which organisations emerge repeatedly and with progressively more complexity. Fontana and Buss do this by using multiple experiments in a reactor with abstract chemical objects.

On one hand, objects with the capacity to reproduce are observed, hypercycles which allow collective auto-replication. When the replication lessens, complex self-maintaining structures emerge. Moreover, it is possible to observe even greater levels of complexity, because structures which are already self-maintaining can combine with others that have similar characteristics and generate new structures with a superior level of organisation. Moreover, as we rise up the hierarchies, we detect the emergence of new laws. It is not difficult to perceive the similarity existing between the objects found on different organisation levels, and their properties, and the varied phenomenology which has gradually appeared over the course of biological evolution: DNA replication, prokaryotic cells, metabolism and the emergence of multicellularity, to name but a few examples.

It was Stephen J. Gould who raised the question of what would happen if we were to press rewind on the film of life, and he placed particular emphasis on the contingent character of what we know about the history of life on this planet. In any case, the fact that it is the first and only "tape" known to us, makes what may appear in the second or any other tape of life unpredictable. It is probable that this statement is not entirely true, because the regularities which appear in the computational experiments of Fontana and Buss, and indeed many others (I have only used a sample of the large amount of research into the matter), emphasise a certain inevitability of emerging regularities and organisational hierarchies.

Perhaps, the best means of combining contingency and necessity in Biology and evolution is through the concept of *open evolution* or *open ended evolution*. Remember that *Life* does not easily capture the open nature of evolution, probably because necessity exerts a great influence on the game in relation to the contingency/necessity pairing. This necessity is strongly linked to the rules determining it, and the consideration, not at all trivial, that no type of mutational variation is taken into account. Only the configuration and placement of the objects on "square one", as well as the size of the aforementioned square, can add a certain historical contingency to the expected result. On the other hand, this does not occur in the reactor with the abstract objects used by Fontana and Buss, where contingency and necessity are more balanced. As such, behaviour that is more akin to evolutionary dynamics can be observed. However, despite the aforementioned balance, which basically involves adding factors that increase the emergence of contingent phenomena, the authors insist on a general conclusion, which is important in relation to this essay. The conclusion involves the possible growth of complexity, linked to the fact that the experiments carried out in the reactor, more or less reflect the accumulative phases of life's evolution, such as replication, unicellularity or multicellularity.

How much necessity must there be in Evolutionary Biology, despite contingency, for it to play a role in determining a certain line of progress? I have explicitly differentiated between closed evolution and open evolution. There is no sense of finality in the evolutionary process; its dynamic is sprinkled with contingent events which can lead it down unforeseen paths, but only to a certain extent. This is the thesis of open evolution supported by Fontana and Buss, given that the primary patterns of Evolutionary Biology and its main characteristics are probably going to repeat themselves in one way or another. I believe that the metaphor of the dinosaurs' intelligence of dinosaurs is a good example, which perfectly combines the relation between necessity and contingency. The path of evolutionary dynamics is precisely that of the emergence of successive complexity following an initial point of simplicity.

## References

Church A (1942) The calculi of lambda conversion. Princeton University Press, Princeton

Fontana W, Buss LW (1994a) The arrival of the fittest: toward a theory of biological organization. Bull Math Biol 56:1–64

Fontana W, Buss LW (1994b) What would be conserved "if the type were played twice". Proc Natl Acad Sci USA 91:757–761

# Part III
# The Cell and Evolution

# Chapter 9
# The Spaces of Evolution

**Abstract** A great problem facing modern research into Biology is to formulate a theory of life that combines the phenotypic and genotypic evolutionary spaces. At the time it was perfectly conceptualised by Lewontin, not without some scepticism about its feasibility and an acknowledgment of the need for detailed knowledge on the laws that combine both spaces. More recently, Wagner argues that such laws are accessible if the studied phenotypes and their associated genotypes can be approached empirically and computationally. To the extent that we can determine such laws we will be able to understand the evolutionary, temporal dynamics thereof.

In his influential book about the nature of genetic change, Lewontin (1974) presents us with a plan which is adequately general in order to outline what the process of life and evolution entail. The first thing to consider is the fact, of great relevance in Biology, that there is a separation between the so-called *genotype and phenotype* spaces. Living beings are not common machines. A trait of a machine is that an algorithm (requiring a finite amount of time to run) can simulate what they do; accordingly we can say that living beings are also machines if we can describe them with an algorithm of this nature. It is not entirely certain that living beings possess this trait, and this is a subject I have considered throughout this book. Notwithstanding, we can continue with a description of life as a *special machine*. When a component breaks down in a washing machine, it does so as it no longer carries out its corresponding function, and normally, the machine ceases to function. It is true that there are machines which are very similar to life, with regard to the fact that like life, they have a peculiar trait whereby even if one of their components fails the machine itself will not stop working. This property is called *resilience*.

However, we should look again at the other key element in the machine of life: the presence of two large components which are delimited and separable, at least conceptually. The first is the information component (genotype) and the other is the working component (phenotype). From this dichotomy arises the essence of living beings' ambivalent capacity to remain the same whilst also gradually changing. This structure with two components, which are the previously mentioned

© The Author(s) 2015
A. Moya, *The Calculus of Life*, SpringerBriefs in Biology,
DOI 10.1007/978-3-319-16970-5_9

spaces, provides living beings with a capacity for resilience, whilst they evolve. This is because the changes taking place in the genotype space ought mostly to be promoters of unviable changes. These mutational changes, or changes of greater importance, ought to produce unviable or unbeneficial changes for the organism. Yet, they do not. It seems that the phenotype space remains unaffected, or only slightly so, if we compare it to what occurs in the genotype space, at least during a certain amount of time and under certain conditions. The fact is that there are connections between both spaces. Thus, if one is altered, this will have some sort of effect on the other. The relationships I refer to are none other than the laws governing the transformation of genotypes into phenotypes and of phenotypes into genotypes. These constitute the huge bundles of laws giving rise to a great amount of current biological research. On one hand, there are those involved in the development of the genotype (ontogeny) and, on the other hand, there are those which have to do with the way in which certain phenotypes can affect derivative genotypes. Both are epigenetic laws.

## 9.1 Laws Governing Organisms

As I have already mentioned, in his book *The Genetic Basis of Evolutionary Change* Lewontin (1974) offered a masterful presentation of a plan which clearly demonstrates the obligatory requirement to investigate all the laws governing the creation of organisms, their ontogeny and eventual population dynamics over time. This plan operates using a structuralist conception of scientific theories; a conception which allows us to capture relatively well the complex phenomenology making up the time variable, as is the case with life and its evolutionary facet. The objective of this chapter is to further explore Lewontin's plan in light of current biological research.

Lewontin had already described the problems faced by Neo-Darwinism with regard to being able to differentiate between the evolutionary forces promoting organic change, which form the core of Population Genetics. At the time, Lewontin was sceptical about the ability to empirically distinguish between the relative values of some forces against others, for example, genetic drift against natural selection. His reasoning was rooted in a merely statistical question regarding sample sizes, in order to effectively differentiate between them. In order to make an adequate comparison, it was necessary to have a high number of specimens and genes, and study the nucleotide sequence of the latter in detail. This outlook has changed considerably, what seemed like an unapproachable task, no longer is, even if it is sometimes a titanic struggle. It is now feasible to carry out research projects to determine the complete genome sequencing of hundreds or thousands of organisms, all depending on how large their respective genomes are. Would Lewontin now be convinced that we find ourselves in a time where it is possible to put forward a complete explanation of the origin and evolution of human beings? Probably not. Accordingly, I must return to the structuralist plan of population dynamics to which I have been referring since the beginning of this chapter.

As mentioned previously, we can distinguish between two large spaces: those of genotypes (G) and of phenotypes (P). Genotype and phenotype spaces are population spaces. That is to say, they are made up of entities which are obviously going to be different from one another as a consequence of the mechanisms generating mutational change or because of genetic variation. The distinction between the spaces is highly important, because in a certain sense, it reveals something unique about what we have come to call the special machine represented by life, and the cell as an archetypal example of it. Its genetic or genotypic domain differs from the phenotypic domain.

Lewontin sets out to explain spaces are made up of different genotypes and phenotypes, where each genotype will give rise to its corresponding phenotype, and vice versa, over the course of biological time. The necessary relations shaping the relationship between both spaces, as well as the spaces themselves, are unavoidably the elements which make up the transformation laws. Specifically:

1. T1: Transformation laws regarding genotypes which become phenotypes.
2. T2: Transformation laws regarding phenotypes which remain as phenotypes.
3. T3: Transformation laws regarding phenotypes which become genotypes.
4. T4: Transformation laws regarding genotypes which remain as genotypes.

Let us say that G1 is the population space of initial genotypes. These genotypes, following certain transformation laws, which we could call epigenetic (T1), allow for the development or creation of corresponding phenotypes or, P1 population. In the same regard, a population of phenotypes, let us call it P2, gives rise to certain genotypes, let us call it G2, returning again to the genotype space. The nature of these epigenetic laws (T3) is one of the major challenges of current biological research. These transformation laws are very important, and not necessarily equal in determining the corresponding phenotype or genotype, depending on the level of complexity of the initial genotypes and phenotypes. The T1 laws are the basis of what is known about mapping the phenotype using the genotype. The T3 laws have to do with the epigenetic modifications involved in shaping certain genotypes. If these laws, as a whole, were strictly linear, then the relationship between the genotypes and phenotypes would be bi-univocal. Namely, given a genotype, we would have a certain phenotype and given a phenotype, we would have a certain genotype. However, if we know anything about the cell, it is that the complex relationship between genotypes and phenotypes is neither bi-univocal nor linear. It is precisely this lack of bi-univocality which makes living machines interesting, to the extent that, as we will see later on, particular failings of a machine, either in genotypic or phenotypic spaces, do not necessarily mean that the organism will fail.

It should be noted that laws remain which have yet to be explained. There are the laws of the phenotype space (T2), where phenotypes interact amongst themselves, when they compete, cooperate, reproduce, etc. These are the laws which give us an account of how some populations change their composition, because factors such as natural selection, differential mating, genetic drift, migration, etc. mean that some phenotypes survive and can give rise to offspring, whilst others cannot. These laws transform the population of phenotypes from P1 into P2. Finally, I have yet to mention the best known laws: those which turned Genetics into a science (T4). These laws are able to explain how an initial population of G2 genotypes, resulting from P2 phenotypes, creates a new generation with new genotype spaces. This occurs by combination, when reproduction is sexual, or by simple descent, when reproduction is asexual or parthenogenetic. This new generation begins a new intergenerational path passing again from genotype to phenotype spaces, and vice versa, with the presence of the different T1–T4 transformation laws.

With this proposal, Lewontin probably formulated the entire series of problems surrounding a causal understanding of all living beings. It is a proposal which defines the existence of the genetic space, information space, and which also describes the space where genotypes develop, the phenotype space. These are not spaces which refer to isolated individuals. Each of them, for example if they are cells, contains its own genetic-information domain, and its functional development, the phenotype domain. It is probable that as Lewontin's proposal emphasises the population aspect, it somewhat neglects the existence of individuals themselves, isolated and locked in their own compartment. Membranes are the compartments in cells; however they are probably implicit in phenotype spaces. The entire plan is very abstract and, to a large extent, devoid of explanatory content in the sense that we still lack a full grasp on the complex range of transformation laws.

Considering that the genotype space is that which gives rise to a new generation, and taking an abstract view of what it means to change from a genotype into a phenotype, and vice versa, assuming bi-univocality in T1 and T3, and paying attention to detail in the T2 and T4 laws, we may be capable of predicting population change over time, and to define evolution as a transformation of genotype spaces. In essence, this is the research topic of Neo-Darwinism. However, we only need to think that this is an approximation, a model which by necessity has certain simple rules T1 and T3. Current biological studies are attempting to show that these rules go beyond mere bi-univocality. Neo-Darwinism is a model of evolutionary change that has neglected T1 and T3. However, this does not mean this model can be deemed a failure. Neglecting T1 and T3 means minimising the development of the phenotype, and ignoring the rules which control the mapping of the genotype changing into a phenotype, and the epigenetic rules of the phenotype becoming a genotype. However, we should remember that the genotype space is the information space, the space where we witness the creation of future generations, and a space which leaves a trace of the past in the genomic record of its components. It is a model that is successful,—indeed, was successful—by concentrating on one area, and which has probably given a better account than any other of the history of life, and the genetic relationships between beings forming the tree of life. Indeed, the current genotype domain bears witness to the past. We also have obvious evidence in the form of fossil records, and we can explain a lot about phenotypic evolutionary change, if we concentrate on the time series of phenotypes in the past. However, the evolutionary history of our genomes is much greater; it takes us back to time immemorial, almost to the dawn of life itself. The greatest contribution made by the study of the aforementioned space, is precisely, the Darwinian confirmation of the tree of life.

## 9.2  Treatable Phenotypes

In his interesting work *The Origins of Evolutionary Innovations*, Wagner (2011) looks at the origin of evolutionary innovations, a concept which is of particular interest when attempting to tackle the series of problems faced due to a limited knowledge of the T1–T4 transformation laws. According to Wagner, we have to look empirically at systems of intermediate complexity which may really allow us to extensively explore the real relationships between genotypes and phenotypes, without the need to make overly simplified assumptions (as Population Genetics would do), or having to use organisms with complex development (which is what happens in the Evolutionary Biology of development or Evo-Devo). Therefore, we must hunt down and capture what we could call *treatable phenotypes*; that is to say, systems of intermediate complexity where we can manage both empirically and computationally a substantial part of possible genotypes and their phenotypes, in order to get a real grasp of transformation laws. If this approximation is feasible, then the laws derived from it will be sufficiently general, both for organisms with simple development, as well as more complex organisms. The fact we possess treatable phenotypes will then make it possible to create what Wagner calls *a theory of evolutionary innovation*. Wagner, like some others, believes that Darwin's theory of evolution falls short when it comes to explaining the emergence of the main evolutionary innovations, an issue Hugo de Vries would call *the arrival of the fittest* in contrast to the Darwinian idea of *the survival of the fittest*. From Wagner's point of view, the aforementioned theory should include an element of *exploration*, according to which even if phenotypes displayed slight adaptation, there would be a possibility for new ones to emerge. In the same regard, such a theory should account for how innovations which have gradually appeared adequately integrated themselves, whilst considering the complications arising from the existence of different levels of biological organisation: namely, an element of *integration (or comprehensiveness)*. Likewise, innovations are largely modular (property of *modularity*). It is somewhat surprising that Wagner mentions the notion of modularity and the integration of modular systems but does not make any reference throughout his theory to the relevance of compositional evolution through symbiosis, that is to say, through the integration of pre-adapted modules coming from different organisms. I briefly looked at this question in Chap. 1. In any case, as previously commented, it does point to the need for the theory to consider combinatory, compositional or modular integrations.

The fact that the genotype-phenotype is not bi-univocal raises, almost by definition, a situation where different genotypes may functionally converge into the same phenotypes. Wagner's theory of innovation must therefore consider *functional convergences* a property which should be explained, in such a way that the same problem can be resolved by different innovations.

The last consideration regarding his theory includes an element of uncertainty which we must consider, as would be the case with any evolutionary theory: I am talking about environmental change. Evidently, this is not a trivial factor, as if innovations are to be tested, the tests must take place in certain environments. Similarly to Watson's compositional theory, Wagner's theory is a non-selectionist theory. Natural selection is, of course, present in order to test innovations; however, we cannot predict the course of environmental changes that would mean the innovations themselves were viable at one given moment and not at another. Environmental change adds an element of uncertainty to the course of evolution. However, it is necessary to remember, and for integration purposes, that historicity or contingency must indeed be considered. Yet, it is also important to consider the circumstances of the necessary emergence of increasing complexity over time.

Therefore, what type of phenotypes will be the most interesting when it comes to developing the aforementioned theory of innovation? Essentially, Wagner resorts back to Lewontin's structuralist plan. The only thing he attempts to do is to give it greater empirical weight, in order to discover the laws of transformation, particularly the epigenetic laws T1 and T3. Wagner maintains that for genotype spaces, we need to be capable of finding out and exploring, empirically and/or computationally, all of the genotypes within a given system. The same should occur with phenotype spaces, we must discover all of them. The most difficult phase will occur when we have to link or find relationships between genotypes and phenotypes, and carry out mapping, knowing that the relationships between them are anything but simple. This is where modern Biology is making great progress, given that there is now a wide range of experimental and computational methods which can be used to determine what type of relationships exist between both spaces and, therefore, facilitate the formulation of the corresponding epigenetic laws. Finally, it is also necessary to point out that in order for the theory to be effective at explaining what occurs in nature, the treatable phenotypes being studied must be contextualised in relation to the population. To a certain extent, Wagner is only emphasising the presence of other laws: those specific to the success of phenotypes (T2) and those relating to the transfer of genetic information to future generations (T4).

In relation to Lewontin's idea, Wagner's proposal is empirical along with the consideration that innovation cannot exclusively originate from gradual evolution. I have already had the opportunity to talk about the latter in previous chapters. However, I have not spoken about what would be examples of treatable phenotypes for Wagner, or intermediate phenotypes. There are three of them: metabolic networks, genetic regulation, and the study of new molecules. It is relatively clear here that we are dealing with systems of intermediate complexity in so much as we can explore a large part of them, if not all of the genotypes, as well as characterise and put the phenotypes into groupings according to the responses observed. Let us consider a particular metabolic network as an example. The genotype space would be formed of the corresponding genes involved, the phenotype space would be made up of the working network itself. The possibility of mapping gene changes in the network is something that can be carried out both computationally, as well as with working experiments evaluating a possible response (synthesis of suitable products, robustness against genetic changes, etc.). On the other hand, there are procedures which allow us to determine the distribution of mutation effects in relation to the biological efficiency of individual carriers. The availability of this type of intermediate phenotype may represent the best option for deriving transformation laws from Lewontin's proposal, particularly epigenetic laws.

# References

Lewontin RC (1974) The genetic bases of evolutionary change. Columbia University Press, New York
Wagner A (2011) The origins of evolutionary innovations. Oxford University Press, Oxford

# Chapter 10
# Computing the Cell

**Abstract** Here I call for the need to revisit cell theory. The idea that the cell is the basic unit of life is well-known and foundational to Biology, but it has not received sufficient attention. We have increasingly detailed knowledge of the intracellular world and all its components, but these are often considered independently. On the other hand, there is excessive theorising about the cell on the basis of its being a black box. The time is ripe to formulate an integrative cell theory.

Biology is said to make sense in the light of evolution (Dobzhansky), but to explain life a theory of the cell is probably more relevant than an evolutionary theory. Evolution is a fundamental part of life, but contrary to what has been assumed about the relevance of different biological theories, we will probably need a cell theory rather than an evolutionary theory to give a full explanation of life. As Barbieri (2003) stated, surprisingly theorising about the cell is not given sufficient consideration, as if we assume we have known all about it since the dawn of Biology when we discovered that organisms are composed of cells (when multicellular), and that life persists through its cells. But the formulation of a theory of the cell means going far beyond these generalisations, however important they may be. Many of the heterodox theorists to whom I refer in this essay have developed models and conceptualisations of the cell that have exceeded the empirical knowledge we need about the cell's interior, its structure, molecules, metabolism, transport, etc. As I have already commented, abstractions are important, but empirical biologists demand detailed knowledge, which nourishes or feeds the knowledge of life. Therefore, here I will discuss in greater depth what the combination of theorising and abstraction of the cell can tell us about its innermost composition in the light of modern Biology.

## 10.1 Cell Theory Revisited

Textbooks on the history of Biology recount the discovery of the cell. Once its presence was established in all kinds of living beings, Theodor Schwann proposed the cell to be the basic unit of life. The origin of life commenced with

© The Author(s) 2015
A. Moya, *The Calculus of Life*, SpringerBriefs in Biology,
DOI 10.1007/978-3-319-16970-5_10

the formation of what could be called a *cellular* entity. A different issue is the fascinating process of chemical synthesis that preceded its conception, and that logically combined three fundamental processes that no living entity, however simple it may be, can do without: membrane, metabolism and genes.

On discovering the cell, when the first microscopic organisms were observed and the cellular structure of complex organisms was spotted, the notion that the cell is the fundamental component of life was taken on board and given the status of a theory and thus became part of the future evolution of science. Notwithstanding the vast heterogeneity of organic beings, they all have one thing in common: their basic fundamental unit is the cell. There are organisms consisting of individual cells that reproduce by partition while others are multicellular and produce offspring through germinal cells. We hardly find any stage in life that is characterised by a state other than cellular. All organisms, absolutely all of them, have cell units in one form or another, which underlie the basis for their existence. Meanwhile, the abovementioned components-membranes, metabolism and genes-perform vital functions, and are used to define a particular entity as *cellular*. One might wonder whether there are machines that could be called *cellular* as they comprise compartments (membranes), metabolism and genes.

The above issue is not trivial. To avoid the cell versus machine conflict, let me put this idea the other way around and say that the cell is a special machine, an idea discussed in a previous chapter. So even if we could formulate an exact definition of these three concepts and find machines to fit such definitions, thus enabling us to state that these machines are cells, I think the opposite would be more interesting: to describe the cell as a type of machine. Indeed, von Neumann and Turing already defined machines exhibiting cell-like behaviour. This issue depends on how far we are willing to admit that a cell, like a machine, can be formulated using a more or less complex algorithm; an algorithm that can be run on hardware, and we should ponder whether it may or may not have a computable property. The eventual computability of the algorithm is equivalent to its ability to solve the problem it should solve. But it is conceivable that such a situation does not arise; that the machine and its corresponding algorithm do not stop, are not computable. If we state that a machine, by definition, can be formulated in terms of a computable algorithm, then it follows that if the cell is a machine it must also be computable, however unusual its *modus operandi*. But is a cell computable? Are all machines computable? Are there machines that are not? If the latter question has an affirmative answer, the cell could be considered a special machine, but not necessarily a computable one. This does not contradict the fact that you may compute the cell, which is the central issue of my thesis. Definitions aside, I maintain that it is interesting to compare the cell to a machine and, in all probability, because it is a very special machine, we must consider that it is not computable, just as that some machines are not. What do we mean when we say a cell is not computable? Essentially this means that its behaviour may be unpredictable; if we start with an algorithm that, in our view, covers each and every one of its properties, we cannot exclude the possibility that it will demonstrate unpredictable behaviour. Such unpredictability is probably crucial and fundamental to life, as it is at the basis of what underpins the actual behaviour of all living things: evolution.

Both Gödel and Turing managed to show properties in formal systems that demonstrated their incompleteness; the impossibility defining true statements with a particular formal system. It is no longer a metaphor that cells can be formulated in formal and computational terms. It seems logical that we can compute a cell, particularly a cell that is simple in terms of compartmentalisation, metabolism and genetics. We are able to do this because currently we can (and increasingly so) list all a cell's components and their interactions, as well as the functionality of its genes and a detailed account of all its proteins and metabolisms. Thus, we have knowledge about all these elements and we know how they interact. This knowledge is like a huge algorithm, whose temporal dynamics can be checked according to the data supplied, in terms of cell maintenance or internal homeostasis, and the products they manufacture. All this information helps us to verify whether the algorithm defines the cell. However, one thing is the algorithm, the ability to reproduce the behaviour of a cell in silico, something we can approximate substantially, but it is quite another question as to whether the cell is computable in the sense I have described. Indeed, it is probably not possible to compute because if it were, we would be ignoring something that is the very essence of life: its ability to evolve. Evolution is probably the indisputable component of the biological world that shows us the impossibility of ever formulating the ultimate cell theory, that is, a complete explanatory theory.

Time exposes the cell to unpredictable changes, which are not exclusively those that may occur in genes as a result of replication errors of the polymerases. There may be other phenomena resulting from the interaction of the three major components of the cellular world, or also those existing between their subcomponents. These interactions would underlie and explain the real and unique phenomena in the cell; phenomena that cannot be explained by a particular algorithm specified by the genes, regardless of how complete this information may be as a reflection of our knowledge of the cell. The major transitions in the story of life fit well within this explanatory scheme. But note that evolution is a by-product of the cell. As we become increasingly more adroit at defining what integrates a live cell, we will also become more adept at formulating and/or tinkering with its behaviour in an infinite game, very akin to the way scientific knowledge is generated, and in sharp contrast to dreams of attaining ultimate theories.

Just as research into the cell has a history, so too does evolution, and both are unfinished, albeit with different non-linear rhythms. Although scientific knowledge is cumulative, its growth is not linear, and it does not move at the same speed in different scientific fields. The discovery of satisfactory explanations or important theories varies widely from one field to another, and some may seem to remain dormant for years when suddenly a breakthrough occurs that changes everything. The history of Molecular Biology is a good example of how the discovery of the structure of DNA was essential to an in-depth understanding of the subcellular world. Meanwhile, scientists were describing the metabolism in great detail, how enzymes and biochemistry operate. Both research areas developed in parallel, but at different rates. Currently we find ourselves at the point of their happy reconciliation. Moreover, new technologies have unlocked the intracellular world and we are now able to scrutinise genes: expression, regulation, metabolites, proteins, kinetics of metabolic pathways, among others. And, therefore, we are in the position to simulate the cell, something that is being done more and more with different approaches in recent decades. This simulation integrates our knowledge of the cell and shows us to what extent the elements incorporated into the simulation reproduce cell behaviour. This is a feedback loop that can obviously provide clues so we can integrate new knowledge into future simulations, and thus we progressively close the gap between the simulated entity and the real entity. Despite this progressive fine-tuning, there will always be doubt surrounding unpredictable behaviour, because if the cell is computable, the algorithm describing it cannot ensure that new properties will not emerge.

The school of Evolutionary Biology resting on the idea that genetics alone holds the key to evolution presents an ambivalent problem because Genetics is able to explain many things, but not all. This fact is somewhat obvious but does not seem to be taken into account much by critics of neo-Darwinism. Focusing on genetics—now on genomes—has provided important results. Most importantly we have verified the tree of life, the relationship between all living beings, and also the existence of evolutionary forces acting on genes and genomes throughout the history of life on our planet. Abstracting from phenotypes, but also considering relevant population-related parameters of informational molecules, such as effective population size, mutation rates or selection coefficients, we are able to account for the evolutionary history of genes and genomes and, in doing so, to reconstruct the tree of life. The tree of life, however, is more than just a tree of genes and genomes. The tree of life also represents cells and organisms and it is necessary, therefore, to have an explanatory theory of the cell and of organisms. Such a theory must necessarily consider the processes that generate evolutionary change giving rise to new cells and organisms.

## 10.2  Organic Memory

In an excellent exercise in Theoretical Biology, Barbieri (2003) argues that the concept of *epigenesis* is critical to understanding life. I have already had occasion to refer to this concept in previous chapters. But Barbieri's notion has an element of generality that should be discussed here. For this author, epigenesis means the succession of new genesis, of new structures or functions. In Lewontin's structuralist proposal for evolution, epigenesis should contemplate the laws of transformation from genotype to phenotype (T1) and from phenotype to genotype (T3). Barbieri goes a little further when he states that the digital information (code) associated with the genetic material is not enough to account for the cellular phenotype. In his view, leaping from one to the other represents bridging a huge gap if we do not contemplate the existence of other *memories* and other *codes* in addition to those associated with DNA and its protein translation. In fact, he even suggests that there are at least five epigenetic processes among genes and cells, namely: those related with the processing of introns to provide messenger RNA; the formation of polypeptides to give the primary protein sequence; those involved in protein folding to provide its tertiary structure, and finally, those involved in protein assembly giving rise to supramolecular complexes or even organelles. We can add three more processes to the five above in order to form organisms, particularly when cells assemble to form tissues, tissues assemble to form organs and, likewise, the organs form organisms. Therefore, if the first step from the gene to the organism is genetic, which involves transforming the information from DNA, by transcription, into messenger RNA, we still have another five of epigenetic nature before reaching the cell phenotype and three more to achieve the organism.

Barbieri suggests that just as there is a code to process the genetic memory, there must be additional codes to process supposed organic memories. Not all cells are equally complex, but he warns us that, for example, bacteria have a code and a memory for signal transduction, and eukaryotes have the same plus intron processing to generate mature messengers. And all these memories, and other possible memories, obviously mean that the information contained in DNA is not sufficient to account for the phenotype. We have to add more memories. We could name them all *the epigenetic memory of the cell* or *molecular epigenesis*, and this highlights the fact that the phenotype is more complex than the genotype. Therefore, one fundamental property of the cell is that it has greater complexity compared to that found in the genetic domain. Barbieri also states that the increase in complexity is *convergent*, considering that the formation of a phenotype requires a sequential coupling of memories. Moreover, if we depart somewhat from the thesis that the lower levels of biological organisation have causal primacy over the higher ones, we can better understand the claim of convergence. Causation may be top down and, therefore, as there is some causal determination of higher levels by the inferior, there may be others acting in the opposite direction. Indeed, reference to multilevel causation is common in the new Systems Biology of cells and organisms (Noble 2011, 2012).

After having defined algorithmic complexity, there is nothing more interesting than being able to apply it not only to genetic memory, but also the rest of memories present in the cell or organisms. Obviously, the algorithmic complexity of a cell would be higher than that of its genetic material, and the algorithms required to compress the respective information would be different in length.

# References

Barbieri M (2003) The organic codes. An introduction to semantic biology. Cambridge University Press, Cambridge

Noble D (2011) Differential and integral views of genetics in computational systems biology. Interface Focus 1:7–15

Noble D (2012) A theory of biological relativity: no privileged level of causation. Interface Focus 2:55–64

# Chapter 11
# Gödel and the Blind Watchmaker

**Abstract** While accepting that contingency is key to biological evolution, we wonder how much need there is for it. It is extremely difficult to talk about trends in evolution, but the fact remains that they are found here and there when evolutionary experiments are repeated. But we should ask, for example, whether there is an unavoidable tendency of life towards progressive complexity. This chapter deals with certain theoretical considerations from Logic and Computing on the conditions necessary to formulate a predictive evolutionary theory.

Biology has historically been caught up in a debate on how to approach living beings: from the analytical or synthetic perspective. The analytical approach encompasses the most successful biological sciences, chiefly Genetics, Molecular Biology, and Evolutionary Biology. They are deemed models of a reductionist approach to appraising living beings because, with few exceptions, the methods and concepts they have developed focus on parts, components, or particular traits. In Genetics the basic features correspond to the Mendelian trait, the mechanics of inheritance, and the laws governing transmission from one generation to the next. Molecular Biology focuses on the chemical nature of genetic traits (genes) and on the molecular machinery involved in their expression. Evolutionary Biology investigates how an organism's fitness is affected by particular genetic traits. Eventually a trait can evolve differentially with respect to any other trait, with or without similar fitness. A gene-centred approach is an accurate description of these three sciences, and during the last 50 years the Biology syllabus has been greatly dominated by this gene-centred analytical view. However, there is more to the analytical approach than this. Analysis means the study of an entity by breaking it down into its parts, and the vast majority of sciences are analytical by definition. Genetics, Molecular Biology, and Evolutionary Biology have developed successful methodological tools to study the genes of living organisms. However, there are many other analytical biological sciences, which approach the living entity by focusing on particular components. It is not unreasonable to argue that the analytical view is a permanent methodological approach to living beings, no matter which organisational or hierarchical level we consider (Ayala 1968).

© The Author(s) 2015
A. Moya, *The Calculus of Life*, SpringerBriefs in Biology,
DOI 10.1007/978-3-319-16970-5_11

Historically, biological sciences approaching living entities analytically have been unequally successful, and it is a fact that those sciences focusing on genes have thrived more than others focusing on other levels of the biological hierarchy. Current Genomic Sciences are the typical by-product of the gene-centred approach to the study of living beings.

But can we approach a living being in a different way? Yes and no. The general basic perception held by many biologists and scientists is that living entities (complex entities, broadly speaking) cannot be appraised via an approach that adds up their parts and, and less so by considering that one single part (for instance, the gene, the genome) is enough to gain sufficient understanding of the living entity as a whole. I would like to point out the difference between analytical and reductionist approaches to science, particularly in Biology. Analytic approaches do not reject that the combination of parts and, therefore, the rules and/or laws derived by working with the parts separately, may eventually come together and help us achieve a better explanation of the living being as a whole. By contrast, the reductionist approach ignores the explanatory relevance of many parts of the living system because it assumes that once we have discovered the laws governing one particular and essential part, they can explain the rest, the whole. It has been argued that the analytical view is a reductionist view of science when, in fact, it is not. The analytical view is probably the most representative of mainstream scientific methodology.

Although these approaches take a different stand on the understanding of complex features, they share a common problem:

How to approach or explain the appearance of emergent properties? For the analytical approach, this question is normally solved a posteriori as follows: Once the emergent property is detected or becomes apparent, it is associated with, or is considered part of, the whole system. This approach suffers from a consistency problem because it does not consider the causal relationship between the corresponding parts. The reductionist approach considers that emergent properties should be explained a priori, in terms of the laws governing the simplest parts of the whole system. Although this approach is consistent, it normally suffers the sufficiency problem because those laws are not available most of the time.

The limitations of both approaches become more obvious when the entity under scientific appraisal is complex, as we realise there are properties of the entity as a whole which, although present, cannot be explained in terms of the rules or laws derived when working with its parts. This perception has been expressed for centuries by many reputed biologists and philosophers, who advocate the synthetic view. As with the use of the term *analysis, synthesis* also has many connotations and it is associated to several research traditions. Some of these traditions have vanished, but others endure. Vitalistic, holistic, and systemic approaches to living beings can be considered synthetic views which, from different scientific backgrounds, question the analytic view's ability to provide a proper description of the living being as a whole. But, what are the scientific achievements of such synthetic approaches? As an expert witness, I would highlight the problems faced by the analytical view when attempting to appraise living beings as a whole; but this discipline is capable of defining new important concepts and/or methods, thus abating the criticisms levelled at the analytic approach.

Analysis is a necessary step in any science, particularly in Biology, and we may wonder about the nature of synthetic inquiry, considering the current status of biological research, particularly at the cellular level. Is there any type of behaviour in the whole system that requires some sort of experimental combinatorial game of the parts to predict and/or to explain it? Moreover, do we possess methods, concepts, and tools to take on such a challenge?

The pursuit of the synthetic approach changes in line with our expanding biological knowledge and current questions differ in nature to the previous ones, probably due to recent and amazing advances in Genomics and Computational Sciences. Now, more than ever before, we can combine many parts of a living being; what is more, we can detect many parts functioning at any given moment and, also, how such parts interact. If we are interested in the interaction between components, it is because many properties of living beings are evidently the by-products of such interactions. One particular but extremely important class of interactions concerns emergent properties. Within the current panorama, the living being is studied via a combination of powerful and successful conceptual and experimental analytical tools.

Such is the case of Systems Biology which enables us to simulate the behaviour of cell systems in silico, and thus provide more and more detailed knowledge. The simulated systems are governed by a set of defined rules (axioms), which can approximate the natural ones gradually but where emergent properties may or may not appear. Let us suppose we can fully mimic any given natural living cell because we have a set of predefined rules and components that enable us to reproduce the properties of the natural one. Such a situation may represent the threshold of our knowledge of a particular living being and, in some way, represents the most ideal approach to acquiring biological knowledge: i.e., a combination of the analytic and synthetic views. Let us define as the final integrative stage, the state of biological knowledge of the particular living being we call a cell. The key questions to be asked regarding such a state of knowledge are as follows: Is the simulated system fully predictable? Is it more predictable that the natural one? Do we really think that any behaviour demonstrated by the natural cell will also occur in the simulated one? The answer to all three questions is "no." The main reasoning behind this answer lies in the Gödel theorems and later derivations in what is known as the Gödel–Turing–Chaitin limit (for a more detailed description, see Moya et al. 2009). I have also referred to these theorems in Chap. 6 and have raised the assumption that material entities, including live ones, can be formulated algorithmically (Penrose 1991).

Applied to any biological system (particularly a cell), Gödel's undecidability theorem states that the cell has properties that are neither provable nor disprovable on the basis of the rules defining the system. In other words, on the basis of the rules and component elements or parts that govern cell behaviour, there might be properties for which it is impossible to tell whether or not they can be derived from the rules of the cell system. Emergent properties are an example of this type of property.

Furthermore, the Gödel theorem of incompleteness states that in a sufficiently well-known cell in which decidability of all properties is required, there will be contradictory properties. The biological translation of that theorem is tremendously important because it states that no matter how well we know a particular cell system, we can find properties and/or behaviours that appear to contradict each other. Contradiction is applied to formal systems and is not a proper empirical description when discussing biological features. Contradiction is the syntactic metaphor for referring to examples of a particular in silico cell showing properties (some of which may be emergent) that contradict another which, like the former, is based on the same operational rules and starting components.

Gödel theorems applied to living cells admit translation within the framework of Turing machines. We can formulate the statement as follows: Functions, structures and properties in general of living cells may appear that cannot be computed by any logical machine. If we consider the cell as a Turing machine (for an extensive review, see Danchin 2009), then a finite procedure (an algorithm) should exist showing us how to compute its behaviour. As stated above, we can visualise an integrative stage of biological knowledge where we can define the rules and components of a living cell. Then, hypothetically we can compute the cell and the corresponding algorithm can be executed using a mechanical calculation device, provided we have unlimited time and storage space. But if Gödel theorems apply to physical and/or biological entities (Penrose 1991), they tell us that we cannot anticipate the emergence of new properties in the cell or the lack of them, and sometimes properties will emerge following contradictory trajectories, no matter how deep our knowledge of the cell may be.

## 11.1   Gödel Incompleteness and Progressive Evolution

As envisaged by Darwin, the history of living beings can be represented in a tree-like form and now we know that a set of events or major transitions have occurred but not necessarily sequentially (i.e., from single replicons to chromosomes, from prokaryotes to eukaryotes, from unicellular to multicellular organisms, etc.), playing an important role in shaping life's diversification. The theory of evolution contemplates the nature of the causative and casual factors able to account not only for those major transitions but also for the astonishing range of biodiversity (species and phylogenetic taxa) and associated extinction events (regular or sporadically dramatic) that our planet has witnessed since life first appeared. Throughout its history, life has displayed myriad emergent properties and, to some extent, life is the perfect model on which to study emergence.

So let us consider the following question: Is there a relationship between the constant appearance of evolutionary novelties and Gödel's theorems? Or, to pose the question in a different way, what is the relationship between Gödel's theorems and evolutionary theory? The neo-Darwinian theory of evolution states that evolution ensues by selecting those genetic variants displaying higher relative fitness.

Random genetic drift can also promote evolutionary change by randomly selecting among genetic variants that are selectively neutral. Evolutionary change is, therefore, governed by these two forces, although others cannot be ruled out. This represents another major debate in the history of Biology (Gould 2002).

Let us imagine the evolutionary process as a kind of executable algorithm that we shall call the *blind watchmaker 1* (BW1) which contemplates all the forces (rules) acting on populations of living and genetically diverse objects. Can we predict the expected outcomes of evolution? As shown in the case of the computation of the cell, the answer would be "no", if we apply Gödel's theorems and assume that, just as when they are applied to formal languages, they also apply to the physical entities (including biological) or materialistic phenomena that can be described algorithmically (Penrose 1991). I am not advocating that evolution is totally unpredictable; it often is for a certain number of situations. But now and again, through evolutionary history, emergent phenomena have appeared. It seems that evolution, emergent phenomena, and the unpredictability of the history of life as a whole are perfectly compatible with Gödel's statements. As beautifully described by Danchin (2009) living systems have an intrinsic ability to constantly create new information to then evolve. This is possible because in the early stages of evolution of life a living device appeared, formed by a unit of coded information (DNA) and another device (the protein machinery) that decodes and recodes the genetic information, which subsequently played an important role in the emergence of other codes and organic memories of an epigenetic nature (Barbieri 2003).

Let us assume that we add new rules to BW1 enabling us to explain that particular phenomenon, which was an emergent phenomenon within BW1. Let us call this new system *Blind Watchmaker 2* (BW2). Although BW2 is more sophisticated and has a wider scope than the former, following Gödel's statements it will be exposed to new unpredictable phenomena. And so on.

Many of the reflections made in this chapter have also received attention from Day (2012) recently. Day's work relates computability, Gödel's incompleteness theorem and the ability to make any kind of prediction about biological evolution. In his theorem, Day demonstrates that assuming the digital nature of inheritance, there must be an inherent limit to the type of questions we can answer about evolution. Expressly he states that unless evolution is progressive, a theory capable of accounting for biological evolution is unfeasible. As expected, his demonstration is similar to Gödel's incompleteness theorem and the halting problem of Turing's theory of computation.

To gauge the scope of Day's theorem, we must previously reflect upon three major issues, namely:

(a)  consider inheritance as digital;
(b)  assume evolution is open-ended;
(c)  determine whether evolution is progressive.

All three issues are complicated, for different reasons. We do not seem to have doubts as to the digital nature of inheritance. The genotype space (remember Lewontin's dynamic structural scheme of evolution discussed in Chap. 9) is where new combinations arise, and of course some of them are transmitted to subsequent generations. But we should also ascertain whether, in addition to DNA, there are other levels of organisation harbouring digital information. This was also demonstrated in the last chapter when briefly discussing Barbieri's Semantic Biology (2003). Taking into account that there is more of one organic code (i.e., other than the genetic code), we must consider their respective memories to, thus, lead to the proper formation of the phenotype. Hence, do these memories support the hypothesis that digitalised information does not reside in the DNA alone but elsewhere? What effect would this supposition have in the potential exploration of genotypes? The answer is that exploration would not take place in that space, but rather in the new spaces corresponding to the respective organic memories. From my point of view, the only thing implied by assuming that more memories exist is that the already vast and infinite world of possible genetic combinations would mushroom, as the potential states of that world would have to be multiplied by each and every one of the feasible states of intermediate epigenetic memories. Although there is also an alternative reflection if we accept that the relationships existing between the hierarchical levels within the cell or the organism are causal and go in both directions (Noble 2012). Accordingly, compared with a mere product of all the states of all the memories, we would have a more limited number of possible states due to causation in both directions, i.e., upward from genotype to the epigenotype and downward from the latter to the genotype. The concept of causation can somehow constrain the generating capacity of new geno-epigenotypes. Indeed, in his work, Day refers to the need to focus more on the evolution of the phenotype than on the genotype. He argues that most likely the same situation of open-ended evolution would exist with a one-to-one genotype-phenotype relationship, i.e., each genotype carries its phenotype, or even with a one-to-many relationship, when each phenotype may be the result of several phenotypes, since the potential phenotypes would require an infinite number of genotypes. Nonetheless, the question remains somewhat open, because if the phenotype space were really limited, or finite, it would not be possible to develop a predictive theory because, as that space would be closed, evolution would not be progressive. But if we combine the states of genetic and epigenetic memories, we then witness a substantial increase in the spectrum of potential final phenotypic states, which takes us to the point on the open-ended nature of evolution. If the genotype space already tends to infinity, the range would probably be even greater if we also consider those related to other memories from the assumption that there is no causal link between them. And therefore, from this perspective, evolution really must be seen as a quest within an infinite geno-epigenotype space.

Day continues his theorem with the proof of the following point: if evolution were open, then we would only be able to formulate a predictive evolutionary theory in the event evolutions were *progressive*. In his demonstration, Day leaves no room for doubt as to the term *progressive*. It means that evolution moves in one direction to find new states that have not occurred previously. He demonstrates this by transforming the genotypic space into a countable set in which each and every genotype can be numbered. Therefore, the set of natural numbers at a given moment in evolutionary time can be related to the set at a later time. Day shows that the number associated with the subsequent state will be higher than the previous state if evolution is progressive. And if that happens, we can make predictions about it. The predictions would be of the following type: "Is it true that a particular phenomenon will occur or, alternatively, it is true that it will not occur."

Recently, Chaitin (2012) reached a similar conclusion on the eventual ability to formulate a predictive theory of evolution, also progressive. However, the respective proposals have a different scope. What is surprising, and greatly so, is the highly convergent way in which Chaitin and Day tackle the demonstration, or at least may surprise scientists working in the fields other than formal and/ or computational sciences. Both authors map the genotype spaces (digital) with natural numbers, although they use different functions to determine the value of said numbers in two consecutive stages. In fact, Day shows that in order to make evolutionary predictions, the integer coding of the state of a population of genotypes must be lower than the state immediately after. Chaitin states that to achieve what he calls *creative evolution*, the search function should be one that gradually approaches, but never reaches it: his famous omega probability in the Turing machine halting problem (Chaitin 2007). As the successive function values increasingly approach the lower edge of this limit, organisms become progressively more effective. Therefore, I conclude that, according to Chaitin, evolution is progressive. The search formula he proposes is well known in the field of computing large natural numbers and is known as the Busy Beaver procedure. Essentially Chaitin tries to formulate a vision of Darwinian evolution by natural selection, and finds that the best mathematical-computational way to reproduce it is by allocating large numbers through this Busy Beaver function. I imagine other forms of creative evolution are not contemplated, and hence the equivalence for Chaitin between Darwinian evolution and progressive evolution.

Day's theorem, in contrast, is more in line with our current knowledge of Evolutionary Biology. The author acknowledges that there are many forms of evolution, such as progressive evolution. And indeed, at certain scales there are records of this. But here we are considering universal evolution. A progressive theory of universal evolution alone would enable us to formulate predictions and, therefore, pose a wholly predictive evolutionary biology. But Day states that in fact we do not know whether large-scale evolution, which contemplates all the changes that have occurred on our planet, has an associated parameter for the increasing development that has occurred during the span of life history and which, in turn, relates to the codes of increasing natural numbers used in his demonstration. In short, Day leaves the question unanswered. It could be argued that this parameter could be some kind of measure of universal biological fitness, of thermodynamic, informational or any other nature. Day argues that to date all endeavours to seek a parameter indicating the progressive nature of evolution are defined retrospectively rather than prospectively. He also argues that we cannot rule out this elusive progress parameter as a highly complex variable of the biological world, which is complicated to interpret.

To summarise the three main points of Day's demonstration (the digital nature of the genotype space, the open-endedness of evolution and the parameter for evolutionary progress), he shows that there are still many uncertainties, particularly of an empirical nature, that would help to more effectively support his demonstration of a predictive theory of evolution. Indeed, we cannot rule out the possibility that there are other digital memories in organisms or that their interaction with DNA creates a finite search space and, therefore, diminishes the possibilities open-ended evolution. But even if such a space were really infinite, we would need to find empirical evidence for a parameter indicator of progressive evolution.

## 11.2   Lessons for Synthetic Biology

Synthetic Biology is a discipline which is currently gaining momentum. Indeed, this field has advanced to such an extent that the prospect of constructing parts of a living entity from molecular components, or a whole entity, looms on the horizon. To go into the definition of a discipline of this nature falls outside the scope of this work; however, it is worth reflecting to what extent we can guarantee the operation of such constructs in the light of uncertainties surrounding our ability to predict the behaviour of a cell or of evolution. It is common knowledge that a quirk of evolutionary thinking is to claim the unpredictability inherent to its history. I have also commented on the possible emergence of properties in those computer systems that try to simulate life, as well as the need to balance contingency and necessity. In the light of these questions, I think we should ponder on the real scope of Synthetic Biology. Currently there are two opposed standpoints from which to view this issue: the engineering approach and the systems approach.

From the engineering standpoint, Synthetic Biology is an engineering discipline (Endy 2005) whereby both the whole cell and any natural or artificial cellular components can be standardised. In fact from this viewpoint, how they behave should be both predictable and controllable. I would like to stress that the emphasis on control is of great relevance because it implies that any man-made biological device should always perform as expected in any suitable environment. In terms of BW1, BW2, etc., this would be a case of a "not-blind engineer-watchmaker" (NBEW).

Although NBEW can predict the outcomes better than BW, we should still consider whether we can rule out Gödel uncertainties. Unlike BW, we can apply two levels of quality control: (1) exploring all imaginable environmental circumstances during the design stage; and (2) when the device is put into a biological chassis and released. It is likely that simple biological devices are much easier to control than living cells and, among cells; it is much easier to control a single minimal cell than a complex one.

But regardless of whether we choose to work on simple devices or on minimal cells, we cannot exclude a priori the appearance of an emergent property and, consequently, we will need to move from NBEW1 to NBEW2 and so on, as explained in the case of natural biological evolution.

We can also consider Synthetic Biology as an applied methodology to create biological systems from which we can gain knowledge. I would call this a "not-blind systems-watchmaker" (NBSW) view. Complementary to NBEW, this view directly embraces complex phenomena and emergent properties. Advances in all areas of Molecular Biology and Computational Biology together with the latest developments in network and graph theory enable us to simulate cellular behaviour, tinker with the cell, and observe the effect of such interventions (Serrano 2007). Although NBSW is also subjected to Gödel's uncertainties, it is conceptually and empirically better prepared to delay the transition from NBSW1 to NBSW2 than NBEW1 to NBEW2. It is simply a matter of gathering all the considerations I have made regarding the empirical evidence needed on the existence of genetic and epigenetic organic memories in the cell and ponder whether there is causation in both directions or, if we talk about evolution, if there is a parameter able to describe the progressive nature of evolution. All uncertainties I discuss here basically support, to some extent, the idea that we should be cautious when considering Synthetic Biology as biological engineering.

# References

Ayala FJ (1968) Biology as an autonomous science. Am Sci 56:207–221
Barbieri M (2003) The organic codes. An introduction to semantic biology. Cambridge University Press, Cambridge
Chaitin G (2007) Meta maths. The quest for Omega. Atlantic Books, London
Chaitin G (2012) Proving Darwin. Making biology mathematical. Pantheon Books, New York
Danchin A (2009) Bacteria as computers making computers. FEMS Microbiol Rev 33:3–26
Day T (2012) Computability, Gödel's incompleteness theorem, and an inherent limit on the predictability of evolution. JR Soc Interface 9:624–639
Endy D (2005) Foundations for engineering biology. Nature 438:449–453
Gould SJ (2002) The structure of evolutionary theory. Belknap Press of Harvard University Press, Cambridge
Moya A, Krasnogor N, Pereto J et al (2009) Goethe's dream: challenges and opportunities for synthetic biology. EMBO Rep 10(Suppl. 1):S28–S32
Noble D (2012) A theory of biological relativity: no privileged level of causation. Interface Focus 2:55–64
Penrose R (1991) La nueva mente del emperador. Biblioteca Mondadori, Madrid
Serrano L (2007) Synthetic biology: promises and challenges (editorial). Molecular Systems Biology 3:158

# Chapter 12
# In Conclusion: Goethe's Dream

**Abstract** Kant deemed a theory of life impossible, while Goethe believed that to understand life we should consider it as a whole. The history of Biology is replete with theoretical confrontations, which—dialectically speaking—are usually resolved by their integration in explanatory theories of greater scope. The major confrontation in Biology is between analytic and synthetic conceptions. Kant and Goethe would probably be astounded to see how the analytical conception has helped provide so many details of the cellular world, sufficient to approach the cell as a whole in the way required by the synthetic tradition. I call this "the realisation of Goethe's dream".

Kant defined living organisms as objects with an intrinsic purpose or value. According to this philosopher organisms are self-organised in such a way that every part is a function of the whole and, likewise, the whole is a function of its parts. Kant had already foreseen the tension between subject and structure, and between the causation exerted by each one upon the other. He also recognised living beings as entities that, due to their extreme complexity, are not amenable to descriptions based on similar fundamental laws to those applied to physics: "There will never be a Newton of a grass blade" he claimed. Less metaphorically, Kant believed that science would not be able to understand living entities by focusing exclusively on their component parts, and, thus, it would never be able to explain a whole organism entirely and comprehensively.

To some extent, the history of biological research could be seen as an attempt to prove Kant wrong on this point. In this final Chapter I will discuss this issue, which has also been touched upon in some of the previous chapters, but here I will do so from a more philosophical and comprehensive perspective.

Historically, two major biology-based strategies have endeavoured to achieve this: *analytical* and *synthetic*. Analytical Biology focuses mainly on the study of the individual components of living organisms. Its latest branch, Molecular Biology, has been particularly successful at producing a deluge of data on the molecular mechanisms underpinning life processes. Indeed these remarkable technological advances help us in our quest to explain life.

© The Author(s) 2015
A. Moya, *The Calculus of Life*, SpringerBriefs in Biology,
DOI 10.1007/978-3-319-16970-5_12

The opposite approach in Biology, the synthetic tradition, states that the essence of a living entity cannot be understood by merely studying its parts. Johann Wolfgang von Goethe, naturalist and writer, advocated this perspective when he asserted that "what a living being is, its essence can be split up into its elements, but it will not be possible to go back and recompose the object to bring it back to life."

Goethe recognised the analytical view's important contribution to our knowledge, but stressed it would not be sufficient to understand life itself. By contrast, the weak point of the synthetic view is that, despite its ontologically correct emphasis on endeavouring to understand a living entity without deconstructing it, this approach lacks a conceptual framework and associated methodological tools that enable the study of living beings as whole entities. In an attempt to give credit to the synthetic approach I have already had occasion to quote Ernst Mayr. This evolutionary biologist justified the importance of notions such as *entelechy* or *élan vital*, which were introduced to overcome the Cartesian interpretation of living beings as machines. Mayr stressed that the interaction among components is as important as the components themselves and, thus, knowledge of both parts and interactions is essential to understanding life.

Also noteworthy is the fact that the conceptual and methodological advances wrought by the analytical view have disclosed a wealth of detailed knowledge of biological parts, resulting in a new synthetic view of biology, known as "synthetic view one" (SV1). This view attempts to synthesise either a protocell (or some of its putative components), or to create a minimal cell based on our knowledge of natural cells. Protocell research focuses on fully synthetic systems such as abiotic chemical components and engineered vesicles that can carry out simple replication or metabolic functions. By contrast, the "top-down" strategy uses information from a natural minimal cell—one with simple genetics and metabolism—to synthesise a new cell. Regardless of the heuristic value of each approach, the second represents the engineering view, which incorporates cutting-edge biotechnology contributing greatly to our knowledge, making it possible to use minimal cells as a chassis to which we can incorporate a set of genes with a specific function. What is more, the said function does not need to have existed previously in nature. SV1 engineering aims to create standardised biological devices with predictable behaviour, regardless of the biological chassis in which they are implemented.

Doubtless, Kant and Goethe would both be amazed by these accomplishments, although these thinkers of centuries past would also feel a touch of pride. They would say: "Yes, we can really make life, but, do we truly understand it, can we explain it?" So, despite the fact we have managed to synthesise life with SV, how much explaining is there left to do? Suffice it to say that—heuristically speaking—the two approaches differ greatly, probably because bottom-up SV1 can better explain what has been achieved than the top-down approach. However, the issue at stake is in achieving this, and although great intellectual effort is required by the first approach it could explain the origin of life on our planet, which is still one of the greatest intellectual challenges of all time. The second approach may have more guarantees of success, but ultimately has less explanatory power.

In addition to the above, there is another approach that is both pragmatic (from an engineering and biotechnological view) and heuristic, which I will call 'synthetic view two' (SV2). Conceptually, this approach is more similar to bottom-up SV1 than top-down and is probably better suited to capture the complexity of biological phenomena. This view considers biological phenomena as systems of interacting components that can be analysed and simulated in increasing detail at both the level of components and the level of the whole system. Furthermore, this view does not exclude emergent properties within the system. This SV approach aims to unite old vitalist aspirations while simultaneously drawing on the conceptual, theoretical and methodological cache of modern Analytical Biology. Thus, in the modern biology scenario, SV2 is a little like the best possible compromise between analytic and synthetic traditions. Keywords of this vision are computational modelling, data systems multi-scale generation and Systems Biology. Let us revisit these terms, starting with computation.

Throughout this book I have repeatedly dealt with the relationship between Computer Science and Biology, and the ability of the former to reproduce the phenomenology of the latter. Computing is an appropriate formal language to describe and reproduce biological phenomena. We realise that this assertion might be seen as controversial because quantitative modelling is highly ambitious, especially as the goal is set perhaps too high given the success that the physical sciences have had in describing the physical world. More specifically, unlike the remarkable success of Calculus as a formal language with which to express, model and derive knowledge about the physical world, Biology has developed no appropriate language so far to formally characterise the complex phenomena of living organisms. I do not wish to imply that we have not tried to apply tools from the realm of Physics in order to model Biology, as we have often done so successfully. For instance, we have the macroscopic continuous and deterministic approach based on ordinary differential equations, which is the most widely used methodology in cellular modelling within Systems and Synthetic Biology. Although differential equations have been successfully applied to different systems, two key assumptions of this approach, namely continuity and determinism, are not always fulfilled. In many cases, the number of particles in the reacting species is low and the reactions are slow. In such cases, the previous assumptions are invalid and mesoscopic; discrete and stochastic approaches, which are closely related to the tools frequently used within computer science, would seem to provide a more suitable alternative.

The latter approaches have been implemented in different computational frameworks ranging from rewriting systems to process algebra. In these computational formalisms, biological entities such as signalling molecules, transcription factors, cell membranes and their interactions are explicitly modelled in a way more similar to reactive systems, rather than by approximating the rate of change for molecular concentrations as in 'calculus-based' modelling. In fact, these modelling approaches are being recognised as perhaps an *executable biological metalanguage*, which could perform detailed and rigorous analysis of biological systems.

Theoretical Computer Science, Logic and Mathematics have profoundly changed the way in which we see the world and our place within it as intelligent beings, and our eventual ability to understand it. Among other works, Gödel's incompleteness theorem, Turing's halting problem, and Chaitin's omega number all demonstrate that there are limits to what can be formally known and proven to be true. Indeed, for any given formal system, there are statements that cannot be shown as either true or false. This undecidability barrier has often been used to demonstrate that a computer cannot be intelligent in the human sense, or to argue that bacteria (Ben-Jacob 1998) and brains (Penrose 1991) must apply some yet unknown physics to perform computations that would apparently operate beyond the Gödel–Turing–Chaiting (GTC) limit. The GTC limit marks the frontier of our knowledge regardless of the amount of time and space we are willing to invest to achieve such knowledge.

The other practical limitation to what we can know is *intractability*. Here we are not talking about a limit to knowledge, but a somewhat less intellectual substance, albeit of practical interest. The fact is, there are certain problems that could be solved in principle but they actually take far too long to solve. Thus unless certain conditions are relaxed such problems remain either unsolved or only partly solved. Certain philosophical arguments consider whether the GTC limit and intractability pose an insurmountable barrier to our knowledge of living beings. In particular, I would like to assess their practical impact on Systems and Synthetic Biology. As way of example, let us consider Synthetic Biology, whose goal is to construct synthetic organisms by applying various modelling and/or engineering techniques. Indeed the so-called white-box models are based on the information that current Biology provides about cell contents and operations at different organisational hierarchies. This was one of the great achievements of the SV1.

We should ask ourselves one fundamental question: "How do we know whether this computational model is the correct one?" Naturally the answer depends on what we mean by "correct". One the one hand, a correct model could be one that is computed, i.e., simulated, quickly enough to be of practical interest, or on the other, one that is slower but more accurate. However, as mentioned above, there are intrinsic limitations to the speed we can achieve, and we must weigh up the price paid for such speed in terms of accuracy. By contrast, it is likely that experimental errors will be larger than those accrued during a simulation; hence, even if, in principle, we could run a simulation on a supercomputer in the future, what would it teach us? We could ask whether a model is reliable, i.e., whether it has been formally verified as accurate. Here again, although great progress has been made in the formal verification of software, there are fundamental limitations of the two previously mentioned types: undecidability and intractability.

This said, intractability has not, historically, been an impenetrable barrier. We are now able to routinely solve specific instances of intractable problems, such as the travelling salesman problem, even for millions of nodes, and approximate the optimal solution. When talking about more or less optimum models to resolve a problem or approximate reality, there is a more insidious slant to the question of whether this is a correct model, i.e., the suitability of the model to capture all our knowledge about a biological system. In other words, how can we compare the behaviour of a computer model with that of a biological system? To address this question, we must, again, make two ontological assumptions.

If we were to assume that, ontologically speaking, biological systems and computer models do not intrinsically differ then they could be compared easily. However, even for systems defined within the same formalism, such as a computer code, there are still fundamental limitations to what we can tell about their behavioural equality. Now, conversely, let us assume that biological and computer systems are intrinsically different. In this case, one might well conclude that there is no efficient method able to assess the adequacy of a computer model to simulate a biological entity. Again, this conclusion might seem premature.

However, we could reject the above extreme assumptions and consider there may be a more realistic methodology able to tackle the adequacy of computational models and biological systems. Harel (2005) proposes a modified Turing-like test for incrementally refining, while simultaneously testing, computer models of biological systems. According to the Harel protocol, we must include in the model all the knowledge we have of a biological system, let us take for instance a model of the nematode *Caenorhabditis elegans*. In this case, it will be deemed successful if, in the eyes of an expert, its behaviour is indistinguishable from that of the actual *C. elegans*. If a given experiment records a discrepancy between the behaviour of the nematode and the model, then this discrepancy should be captured in a new version of the model, and thus models will improve gradually over time. Indeed, this is the very essence of Systems Biology. The proposed methodology does not rest on an assumption as to whether living organisms can be accurately modelled on a computer; hence, in principle, it is a valid protocol to follow in order to improve systems-biology models, regardless of any putative fundamental limitation dictated by the GTC limit or intractability limit.

In a similar vein, we have the proposal by Cronin et al. (2006). These authors were pursuing the same goal as Harel but within the context of Synthetic rather than Systems Biology, and proposed the use of real cells within a Turing-like test to measure whether artificially engineered protocells behave similarly to their authentic counterparts. Interestingly, this would be both a practical and a valid test for synthetic life, even if the second ontological assumption, i.e., that computer and biological systems are fundamentally different, were to be proven true. Indeed, here, *nature* itself would be doing the *simulation* instead of a computer program as in the Harel proposal.

Among the most recent studies on the computation of the cell we find the work by Karr et al. (2012), who made a computational model of the human pathogen *Mycoplasma genitalium*. Interestingly, their model is able to predict the phenotype of this bacterium from its genotype. However, on delving into the finer details of the procedure, we quickly realise that the model does not incorporate genotypic information alone, but also other intermediate organic memories in the epigenetic hierarchy (Barbieri 2003).

When we consider long-term evolution, however, and how some other entities give rise to others over geologic time, we must pay greater attention to the eventual limits of GTC when it comes to emergent properties. Indeed, GTC limits and intractability lead to practical emergent properties when modelling and implementing synthetic systems (let us remember the case of Fontana and Buss' algorithmic Chemistry).

The concept of *emergent property* is fundamental to Biology: most biological characteristics and evolutionary advances are not predictable, owing either to intractability or to formal undecidability. In short, emergent phenomena will appear, sooner rather than later, within any sufficiently complex system. This fact calls for rethinking the axiomatic bases and methods used in modelling, in such a way that the new biological system would include the old rules but also add new rules integrating the emergent features.

But on a shorter time scale, when it comes to constructing life, I do not think we should ignore the above limits. Synthetic Biology has been interpreted as an engineering discipline, whereby both the whole cell and any natural or artificial cellular components should be standardised, i.e., they should demonstrate predictable and controllable behaviour. The emphasis on control is of great bearing because it means that any man-made biological device should always behave as expected in any suitable environment. To avoid uncertainties, we need to apply two levels of quality control as I explained in the previous chapter. The first in the design phase, would explore all imaginable environmental circumstances; and the second would be implemented when the device is put into a biological chassis and released. In these cases, can we exclude the occurrence of a new property? Synthetic Biology can also be regarded as an applied method to construct biological systems from which we can gain knowledge. Complementary to the engineering perspective, this approach embraces complex phenomena and emergent properties. This view, referred to as the SV2 approach to Synthetic Biology is probably better suited to meet the expectations of biologists.

The history of Biology could be interpreted as an ongoing dialectic battle between two opposing views on how to deal with and comprehend living entities: integrative versus analytical. The former promotes a cohesive approach, whereby parts are integrated into wholes, as the only way to comprehend the true nature of a living entity; its best exponent is SV2 or modern Systems Biology. The second view approaches life by concentrating mainly on its components, as in SV1. This analytical approach gathers huge amounts of data from all levels of the biological hierarchy, but in particular the cell, as well as the complex interactions taking place. The reductionist approach, SV1, suggests that emergent features are inherently present in the properties of the most basic parts. This aside, the latter analytical approach has led to the development of powerful conceptual and methodological tools, enabling us to dissect and scrutinise the components of living things in such a way that it has led to SV2 and, in my view, the realisation of Goethe's dream.

Synthetic Biology is, or should be, a happy blend of both traditions, and constitutes an example of what might be beyond that dream. Scientists are convinced of the feasibility of building a living being, or part of one, whether it is one that already exists or a purely man-made entity. Advances of this type in Biology require serious reflection on the responsibilities involved, and on the consequences that may result from such tinkering (Moya 2011). Indeed, humankind should ponder upon the myriad paths that such developments may lead us down in the not too distant future.

# References

Barbieri M (2003) The organic codes: an introduction to semantic biology. Cambridge University Press, Cambridge

Ben-Jacob E (1998) Bacterial wisdom, Godel's theorem and creative genomic webs. Phys A 48:57–76

Cronin L, Krasnogor N, Davis BG et al (2006) The imitation game—a computational chemical approach to recognizing life. Nat Biotechnol 24:1203–1206

Harel D (2005) A Turing-like test for biological modeling. Nat Biotechnol 23:495–496

Karr JR, Sanghvi JC, Macklin DN et al (2012) A whole-cell computational model predicts phenotype from genotype. Cell 150:389–401

Moya A (2011) Naturaleza y futuro del hombre. Editorial Síntesis, Madrid

Penrose R (1991) La nueva mente del emperador. Biblioteca Mondadori, Madrid

# Index

© The Author(s) 2015
A. Moya, *The Calculus of Life*, SpringerBriefs in Biology,
DOI 10.1007/978-3-319-16970-5

Printed in the United States
By Bookmasters